Wireless Power Transfer via Radiowaves

Wireless Power Transfer via Radiowaves

Naoki Shinohara

Series Editor
Pierre-Noël Favennec

ISTE

WILEY

First published 2014 in Great Britain and the United States by ISTE Ltd and John Wiley & Sons, Inc.

ISTE Ltd
27-37 St George's Road
London SW19 4EU
UK

www.iste.co.uk

John Wiley & Sons, Inc.
111 River Street
Hoboken, NJ 07030
USA

www.wiley.com

Library of Congress Control Number: 2013952542

British Library Cataloguing-in-Publication Data
A CIP record for this book is available from the British Library
ISBN 978-1-84821-605-1

Table of Contents

Introduction

Wireless power transfer (WPT) is a promising technology based on electromagnetic theory and radiowave theory, representing the combined application of electrical and radio sciences. There are numerous WPT technologies, such as inductive coupling WPT (Figure I.1(a)) and resonant coupling WPT (Figure I.1(b)) as short distance WPT. In addition, WPT via radiowaves has been developed as a long-distance WPT technology, which includes focused beam microwave power transfer (MPT) (Figure I.1(c)) and energy harvesting from broadcasted radiowaves or diffused wireless power (Figure I.1(d)). Both inductive coupling WPT and resonant coupling WPT are based on electromagnetic theory. A transmitter and a receiver are electromagnetically coupled and power is wirelessly transmitted via an electric field, a magnetic field, or an electromagnetic field. Unlike the short-range technologies, WPT via radiowaves does not require coupling between the transmitter and the receiver, but it uses radiated electromagnetic waves. WPT via radiowaves requires higher frequencies, such as microwaves, to focus on the wireless power effectively. The general characteristics of various WPT technologies are described in Table I.1. The primary difference between inductive and resonant coupling is the presence of non-resonance and resonance, respectively.

In the case of short distances, inductive or resonant coupling technologies using coils are effective, but the coupling distance is limited by coupling theory even for resonant coupling. To expand the distance over which power can be transferred, radiowaves are necessary to carry the energy. For WPT via radiowaves, antennas are used for power radiation and receiving. The antenna serves as a resonator to radiate radiowaves effectively.

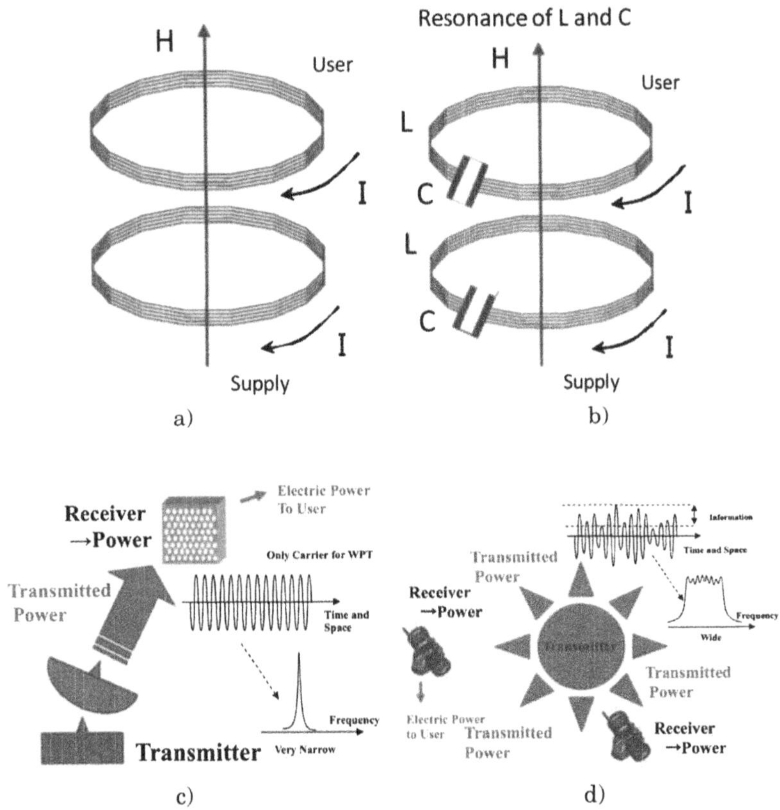

Figure I.1. *Various WPT technologies: a) inductive coupling, b) resonant coupling, c) WPT via focused beam radiowaves and d) energy harvesting via diffused radiowaves*

	WPT via radiowaves	Resonant coupling	Inductive coupling
Field	Electromagnetic (EM)	Resonance (electric, magnetic, or EM)	Magnetic field
Method	Antenna	Resonator	Coil
Efficiency	Low to high	High	High
Distance	Short to long	Medium	Short
Power	Low to high	High	High
Safety	EM	Under discussion (Evanescent)	Magnetic
Regulation	Radiowave	Under discussion	Under discussion

Table I.1. *Characteristics of WPT technologies*

All WPT technologies are based on Maxwell's equations. However, there are minor differences in their applications. It is now possible to use higher frequency radiowaves, for example in the gigahertz (GHz) range of microwaves, to focus on electric power wirelessly at sufficient levels and on beam efficiency to satisfy the needs of many applications. A WPT system is shown in Figure I.2. Frequency converters are used to transform electricity into wireless power, and vice versa. The primary difference between electricity and wireless power is only a matter of the frequency. It is also possible to transmit wireless power from multitransmitting antennas to multireceiving antennas like broadcasting and wireless communications because the antennas are not coupled electromagnetically. Very low power WPT systems do not even require battery storage and can be run on the energy harvested from ambient radio frequency (RF) and microwave radiation. Additionally, the extent of battery storage can be reduced when wireless power is widespread because batteries can be charged wirelessly and, hence, shortages of battery storage are not a concern.

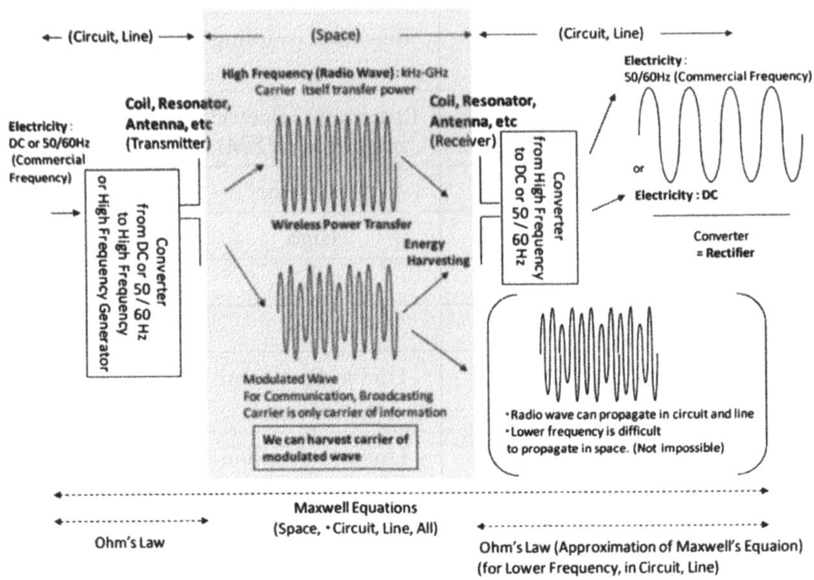

Figure I.2. *A WPT system*

This scenario will soon be a reality. WPT via radiowaves is, in fact, a valuable and convenient technology that can be used to charge batteries in mobile phones, notebook personal computers (PCs), electric vehicles (EVs), as well as battery storage for light emitting diodes (LEDs), integrated circuits (ICs) and other equipment. In the near future, stable CO_2-free electric power from space will be possible using WPT technology. The concept of a space-based solar power satellite (SPS) is supported by MPT. The SPS is a gigantic power station built in space. The SPS is a primary system of significance to the concept of the "humanosphere". The humanosphere concept corresponds to the description of Earth as a space where human activity takes place. Humanospheric science is defined as an interdisciplinary science that conducts research concerning the humanosphere. This science is a significant development for the future of the human race. A pictographic description of the humanosphere and humanospheric science is shown in

Figure I.3. To maintain human welfare and the current standard of living, or even to avoid disaster during this century, the issues of energy, food and the environment should be seriously addressed. Presently, the increasing demand for electricity conflicts with the demand for a clean environment due to the use of fossil-based power production. Electricity has heretofore been generated through various methods, such as hydroelectric power, fossil thermal power and atomic power. However, some of these methods are causes of environmental and pollution issues all over the world. Under these circumstances, research has been carried out to investigate the possibility of building power stations in space to transmit electricity, generated in space, to the Earth via radiowaves. It is believed that WPT technologies support this vision of a bright future.

Figure I.3. *A pictographic description of the "humanosphere" and humanospheric science*

Chapter 1

History, Present and Future of WPT

1.1. Theoretical predictions and the first trial in the 19th Century

In 1864, James C. Maxwell predicted the existence of radiowaves by means of a mathematical model. The so-called Maxwell equations are the most famous and most successful formulas. In 1884, John H. Poynting realized that the Poynting vector would play an important role in quantifying electromagnetic energy. In 1888, bolstered by Maxwell's theory, Heinrich Hertz first succeeded in showing experimental evidence of radiowaves using his spark-gap radio transmitter. The prediction and evidence of radiowaves toward the end of the 19th Century was the beginning of wireless power transfer (WPT).

During the same period, when Marchese G. Marconi and Reginald Fessenden pioneered communication via radiowaves, Nicola Tesla suggested the idea of wireless power transfer and carried out the first WPT experiments in 1899 [TES 04a, TES 04b]. He said "This energy will be collected all over the globe preferably in small amounts, ranging from a fraction of one to a few horse-power. One of its chief uses will be the illumination of isolated homes". Tesla actually built a gigantic coil that was connected to a 200 ft high mast with a 3 ft diameter ball at its top. The device was called the "Tesla Tower" (Figure 1.1). Tesla fed 300 kw of power to the coil that resonated at a frequency of 150 kHz. The radio frequency (RF) potential at the top sphere reached 100 MV. Unfortunately, the experiment failed because the transmitted power was diffused in all directions using 150 kHz radiowaves, whose wavelength was 21 km. After this first WPT trial, the history of radiowaves has been dominated by wireless communications and remote sensing.

Figure 1.1. *The Tesla tower*

1.2. Rejuvenated WPT by microwaves in the 1960s

To focus on the transmitted power and to increase the transfer efficiency, a higher frequency than that used by Tesla is required. In the 1930s, a great deal of progress in generating high-power microwaves in the 1–10 GHz range was achieved by the invention of the magnetron and the klystron. After World War II, high-power and high-efficiency microwave tubes were advanced by the development of radar technology. The power delivered to a receiver can be concentrated with microwaves. WPT using microwaves is called microwave power transfer (MPT).

On the basis of development of microwave tubes during World War II, W.C. Brown introduced the first MPT research and development in the 1960s. First, Brown developed a rectifying antenna, which he named a "rectenna" for receiving and rectifying microwaves. The efficiency of the first rectenna developed in 1963 was 50% at an output of 4 WDC and 40% at an output of 7 WDC, respectively [BRO 84]. With the rectenna, Brown successfully applied MPT to a wired helicopter in 1964 and to a free-flying helicopter in 1968 (Figure 1.2). In the 1970s, Brown attempted to increase the total DC–RF–transfer–RF–DC efficiency using 2.45 GHz microwaves. The overall DC–DC efficiency was only 26.5% at an output of 39 WDC in the Marshall Space Flight Center tests of 1970 [BRO 73a]. In 1975, the overall DC–DC efficiency finally attained 54% at an output of 495 WDC using the Raytheon Laboratory magnetron (Figure 1.3) [BRO 84]. In parallel, Brown, Richard Dickinson and his team succeeded in the largest MPT demonstration up to that time in 1975 at the Venus Site of the JPL Goldstone Facility (Figure 1.4). The distance between the transmitting parabolic antenna, whose diameter was 26 m, and a rectenna array, whose size was 3.4 m × 7.2 m, was 1 mile. The transmitted 2.388 GHz microwave signal was 450 kW from the klystron and the rectified DC power achieved was 30 kW DC with a 82.5% rectifying efficiency. On the basis of

Brown's work, P.E. Glaser proposed a solar power satellite (SPS) system in 1968 [GLA 68].

Figure 1.2. *MPT Helicopter demonstration by W.C. Brown in 1964 [BRO 84]*

Figure 1.3. *MPT laboratory experiment by W.C. Brown in 1975 [BRO 84]*

Figure 1.4. *First ground-to-ground MPT experiment in 1975 at the Venus Site of the JPL Goldstone Facility*

But after the MPT experiments of the 1960s, SPS applications have led the field in MPT research [MCS 02, MAT 02a]. Because of the theoretical calculation, the large antenna size required to achieve high-beam efficiency to a far distant target, an MPT system designed for SPS did not seem to be suitable for commercial applications. However, even if the antenna size was to become larger, there would be the other merits of the space-based SPS. The SPS is designed as a huge SPS in geostationary orbit, 36,000 km above the Earth's surface, where there is no cloud cover and no night throughout the year. Microwave energy is not absorbed by air, cloud and rain; therefore, it is possible to obtain approximately 10 times the solar power, a stable and CO_2-free energy source, from the SPS using MPT technology than that from terrestrial solar sources. As a result of the high benefits expected, MPT research mainly focused on SPS applications during the late 20th Century.

Numerous Japanese scientists developed MPT technologies and research throughout the 1980s [MAT 95a, MAT 02a]. In 1983 and 1993, Hiroshi Matsumoto's team carried out the first MPT experiment in space. The rocket experiment in 1983 was called the microwave ionosphere nonlinear interaction experiment (MINIX) (Figure 1.5), and International Space Year – Microwave Energy Transmission in Space (ISY-METS) was conducted in 1993. These experiments focused on the nonlinear interaction between intense microwaves and ionospheric plasmas. In the MINIX experiment, the researchers used a cooker-type 800 W, 2.45 GHz magnetron for a microwave transmitter. New wave–particle interaction phenomena were observed during the MINIX study. Plasma theory and computer experiments supported the observations [MAT 95b, MAT 95c]. These rocket experiments were directed toward SPS applications.

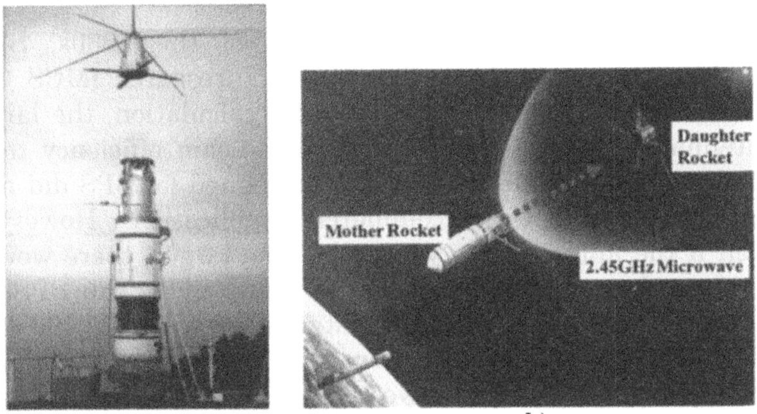

a) b)

Figure 1.5. *The first rocket experiment by Matsumoto in Japan in 1983, called the MINIX project: a) image of mother and daughter rockets; b) image of the experiment*

During the 1990s, numerous MPT laboratory and field experiments were carried out all over the world. This research was not only for SPS but also for other commercial MPT applications. Researchers often used 2.45 or 5.8 GHz frequencies of the industry, science and medical (ISM) band for MPT systems. A Canadian group of the Communication Research Centre (CRC) successfully conducted a fuel-free airplane flight experiment using MPT in 1987, which was called stationary high-altitude relay platform (SHARP) (Figure 1.6) [SCH 88, SHA 88]. They transmitted a 2.45 GHz, 10 kW microwave signal to a model airplane, having a total length of 2.9 m and a wing span of 4.5 m, flying more than 150 m above ground level. In the United States, a great deal of MPT research and development continued after Brown. For instance, retrodirective microwave transmitters, rectennas, new devices and microwave circuit technologies were investigated [BRO 88]. In Japan, several field MPT experiments were conducted, such as fuel-free airplane flight experiments with MPT phased arrays operating at 2.411 GHz for the microwave lifted airplane experiment (MILAX) project in 1992 (Figure 1.7) [MAT 93], ground-to-ground MPT experiments operating at 2.45 GHz were conducted by power companies and universities in 1994–1995 (Figure 1.8) [SHI 98a], and fuel-free light airship experiments using MPT operating at 2.45 GHz in 1995 [KAY 96]. The target system used in the MILAX project was the Japanese SHARP. Kobe University and Communications Research Laboratory (CRL; present National Institute of Information and Communications Technology (NICT)) group in Japan succeeded in an MPT field experiment involving a flying airship in 1995. They called it the Energy Transmission toward High-altitude long endurance airship Experiment (ETHER) project. This research group transmitted 2.45 GHz, 10 kW microwaves to a flying airship 35–45 m above ground level. In these experiments, except in those of the MILAX project, researchers adopted a parabolic antenna MPT system using a microwave tube. A phased array system was used only in the MILAX project, and this project was the first MPT field experiment to use it. In parallel with developments in Japan, varieties of microwave transmitters, retrodirective microwave transmitters and, especially, rectennas were also developed. In Europe, some unique technologies are presently being developed. Researchers had planned ground-to-ground MPT experiments on Réunion Island (Figure 1.9) [CEL 97, CEL 04], but the project has not yet been carried out.

Figure 1.6. *The Canadian SHARP flight experiment and the 1/8 model airplane in 1987 [SHA 88]*

Figure 1.7. *The MILAX project airplane experiment showing the model airplane and the phased array used in Japan in 1992*

Figure 1.8. *Ground-to-ground MPT experiment in Japan in 1994–1995*

Figure 1.9. *Grand bassin, Réunion, France and their prototype rectenna [CEL 04]*

1.3. Inductive coupling WPT projects in the 20th Century

Maxwell integrated Ampere's law and Faraday's law into his equations. Prior to Maxwell's equations, it was known from Ampere's Law and Faraday's Law that a current creates a magnetic field and a changing magnetic field recreates a current. Two conductors are referred to as mutual-inductively coupled or magnetically coupled when they are configured such that a change in the current flow through one conductor induces a voltage across the ends of the other via electromagnetic induction. The phenomenon is called inductive coupling and is applied for power generators and transformers. Contrary to WPT via radiowaves, lower frequencies in the kilohertz to megahertz range are typically used for inductive coupling WPT.

Parallel to Tesla's first WPT experiments, M. Hutin and M. Le-Blanc proposed an apparatus and method for powering an electrical vehicle (EV) inductively in 1894 using an approximately 3 kHz AC generator [HUT 94]. EVs were developed shortly after the development of the steam engine, approximately 100 years ago. However, the EV became less popular with development of the internal combustion engine. As a result, after Hutin and Le-Blanc, the EV inductive coupling WPT charger was forgotten like Tesla's dream of WPT.

Professor Don Otto of the University of Auckland in New Zealand proposed an inductively powered vehicle in 1972 using the power generated at 10 kHz, by a force commutated sinusoidal silicon controlled rectifier inverter [OTT 74]. He adopted two circular cross-section conductors of copper, one of which was placed on the road as a transmitter and the other was placed on the body of the EV at a position 20 cm above the road surface. In 1978, the USA group of J.G. Bolger, F.A. Kirsten and S. Ng carried out the first EV-WPT application in the United States. They used a 180 Hz and 20 kW WPT system whose size was 60 cm × 1.52 m and its air gap was 2.5 cm (Figure 1.10). In the 1980s, another WPT project was conducted in California, USA. It was called the Partner for Advanced Transit and Highways (PATH) project (Figure 1.11) [BOL 78, ZEL 82]. The work achieved a 60% efficiency powering a bus using machine-generated electric power at 400 Hz coupled across an air gap to drive the bus at a distance of 50–100 m [SYS 94]. Researchers in France and Germany conducted EV-WPT projects in the 1980s–1990s known as TULIP (*Transport Urban, Individuel et Public – Public and Individual Urban Transport)* and Inductive Power Transfer (IPT), respectively.

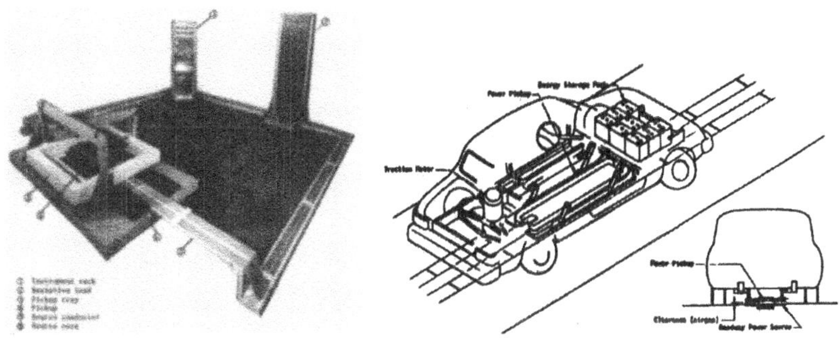

a) b)

Figure 1.10. *a) Experimental setup of the first U.S. EV-WPT experiment; b) image of the experiment*

Inductive coupling is not only used for EV-WPT research projects but also for commercial, non-attached power supply of an electric kettle, an electric shaver, an electric toothbrush, etc. The integrated circuit (IC) card is one of the most famous applications of inductive coupling. In 1995, Sony proposed and developed the Felica system in Japan that used the IC card and adopted an inductive coupling wireless power supply with a 13.56 MHz signal. These inductive coupling WPT applications are used globally. However, there has yet to be any global standard developed for WPT technology and WPT systems vary considerably.

Figure 1.11. *Experimental setup of the PATH project in the U.S. in the 1980s*

1.4. WPT as a game-changing technology in the 21st Century

The other recent trend in WPT began with the use of resonant coupling in the United States by the Massachusetts Institute of Technology (MIT) [KUR 07] in 2006. The resonant coupler is well known as a microwave band-pass filter (BPF). The team at MIT applied the BPF to wireless power transfer. With this technique, a large amount of power (from watts to kilowatts) can be transmitted without radiation over mid-length distances (more than a few meters) at low frequencies (less than 10 MHz) using simple resonant circuits. It became evident that resonant coupling WPT is more suitable for commercial needs. Resonant coupling is based on the inductive coupling of magnetic fields. As shown in section 1.3, various research projects and products using inductive coupling WPT technology have been pursued prior to MIT's revolution. This research and resulting products have supported both theory and development of resonant coupling WPT technology.

The Wireless Power Consortium (WPC) was established in December 2008 to develop wireless charging of mobile phones. They established the "Qi" standard for inductive coupling WPT in December 2010. Although there were some wireless charger developments for mobile phones with inductive coupling WPT, the work did not bear any commercially successful products. The "Qi" standard spread quickly in the world and wireless chargers using this standard have been able to be purchased of late. In addition to the WPC, various alliances, consortiums and forums for establishing a WPT standard and for promoting commercial applications have arisen as follows:

– WPC that established the "Qi" standard for inductive coupling WPT (established in December 2008, 166 member companies) [WPC 13].

– Alliance for Wireless Power (A4WP) to promote resonant coupling WPT (established in May 2012, 60 member companies) [A4W 13].

– Power Matters Alliance (PMA) to promote inductive coupling WPT, which is supported by IEEE-SA (established in March 2012, 95 member companies) [PMA 13].

– Broadband Wireless Forum (BWF) in Japan (established in July 2009, 55 member companies) [BWF 13].

– Wireless Power Consortium for Practical Applications (WiPoT) in Japan for MPT and other WPT technologies (established in April 2013, 28 member companies, 30 universities) [WIP 13].

– Wireless Power Management Consortium (WPMc) in Japan for direct current resonant coupling (established in July 2013, 21 member companies) [WPM 13].

– Energy Harvesting Consortium (EHC) in Japan for energy harvesting, which includes power generation from vibration, heat, light and radiowaves (established in May 2010, 60 member companies) [EHC 13].

– Korean Wireless Power Forum (KWPF) (established in 2011) [KWP 13].

(Data as of September 2013)

The initial targets of the above forums and alliances are wireless charging of mobile phones and/or in-house applications of WPT. These groups aim to establish a global standard for commercial WPT and to expand commercial WPT applications in the world.

Resonant coupling WPT not only represents a coil system in a magnetic field but also an electrostatic induction coupling system with coupled conductors. In Japan, a wireless charger using resonant electrostatic induction coupling WPT technology has been commercialized for tablet charging. Additional research in Japan has been directed toward resonant electrostatic induction coupling WPT, which will be applied to free positioning electrical points [HAR 11]. This application uses 600 kHz, 1.2 MHz and 2.0 MHz signals selectively and has succeeded in 150 W of wireless power transfer with over 95% efficiency.

Many other high-power WPT activities exist. As an example, inductive coupling WPT has been applied to wireless charging of an electric bus by Hino Motors Ltd. [HIN 13] and Showa Aircraft Industry [SHO 13] in Japan. This so-called "inductive power transfer" was demonstrated and an inductive power transfer hybrid system was operated from April 13 to 27, 2009 (Figure 1.12).

There have been some studies on the application of resonant coupling WPT for electric vehicles [IMU 09, IMU 10]. Toyohashi University of Technology in Japan proposed a new concept of power transfer through a capacitor composed of a steel belt attached to the inside of a tire and a metal plate attached to the road (Figure 1.13) [HAN 11]. In 2011, Toyota Motor Co. invested in WiTricity Co., which is one of the first commercial developers of resonant coupling for wireless power transmission. IHI Co. was given a license from WiTricity Co. in 2011. Several other Japanese EV-WPT activities are described elsewhere [SHI 13a].

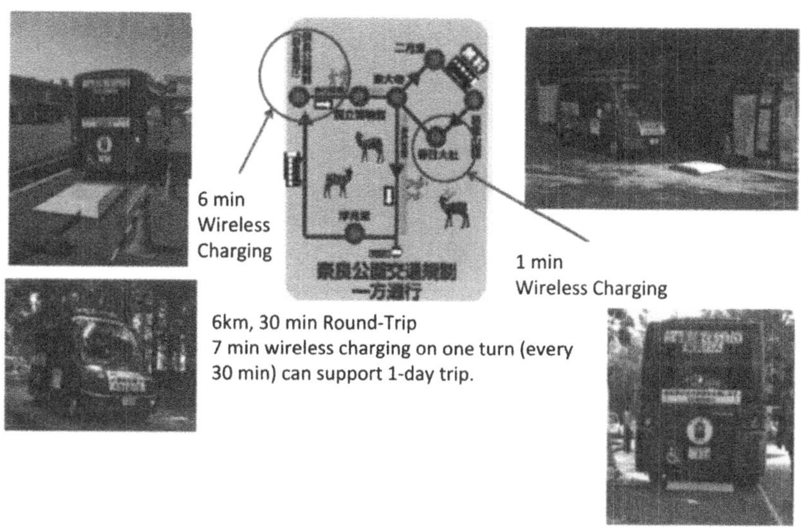

Figure 1.12. *Field experimental setup for EV-WPT conducted in Japan in 2009*

In Korea, a wireless power supply using a resonant coupling technique has been applied to an online electric vehicle (OLEV) (Figure 1.14) [AHN 11]. Power from the 60 Hz supply is converted to a frequency of 20 kHz by an inverter stage. A total of 60 kW of power may be transferred wirelessly from power lines with 80% efficiency.

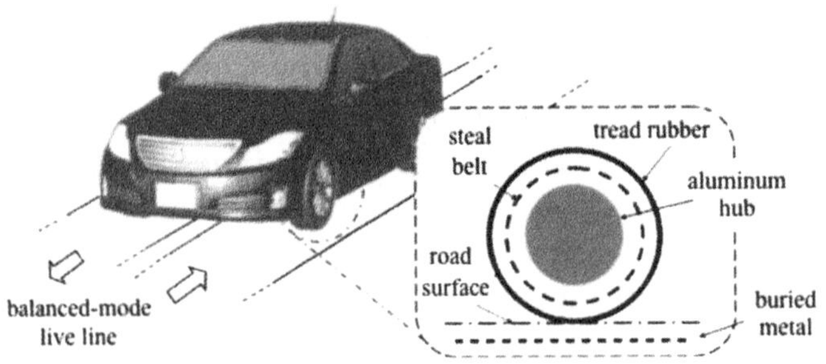

a)

b)

Figure 1.13. *a) Diagram of the power transfer system proposed by Toyohashi University of Technology, Japan [HAN 11]; b) experimental setup*

Figure 1.14. *Wireless power supply for an online electric vehicle (OLEV) in Korea [AHN 11]*

In the early 21st Century, G. Covic's group in New Zealand continued to develop EV-WPT systems with inductive coupling. The group conducted research on a system composed of a multiphase track with multicoil receivers in 2003–2005, and IPT road pads with multicoil receivers in 2009–2010 [COV 12]. The German company Bombardier developed 250 kW WPT railways until 2012 [MEI 12]. Bombardier, BS-Verkehrs-AG, BS-energy and TU-Braunschweig began an EV-WPT project called Elecrobus in Branschweig in May 2012 [MEI 12]. It is a field EV-WPT experiment in downtown Branschweig.

In addition to these EV-WPT research activities, some companies, for example Evatran Co., Audi Co., SEW-EURODRIVE Co. and UniServices Co., introduced inductive coupling wireless power transmission systems for EV applications as commercial products. HaloIPT/Qualcomm Co. adopted resonant coupling WPT and began field experiments in London in 2012. Evatran Co. introduced a plugless level 2 EV Charging System using inductive coupling technology as a commercial product in 2013 [PLU 13]. To date, various WPT systems for robots and vehicles based on inductive coupling WPT technology have been installed in factories all over the world.

Contrary to inductive coupling and resonance coupling WPT, MPT applications seem to have been delayed because of radiowave regulation problems. Inductive coupling and resonance coupling produce less radiation as compared to MPT systems. However, MPT research groups have proposed advanced MPT systems as being acceptable for commercial use.

One such advanced MPT system is the phased array system. The phased array, with which a highly directional power beam can be formed and controlled, is an important aspect in the development of a practical MPT system. The phased array is typically used for radar or remote sensing. For example, a phased array in the S-band, which was composed of 4,000 phased shifters/array and 936,000 manufactured elements, is used in the AEGIS radar system [BRO 06]. A Japanese group, whose members come from the National Institute of Polar Research, the University of Tokyo, Kyoto University, many universities, national institutes and companies, is building a mesosphere/stratosphere/troposphere (MST) radar/incoherent scatter (IS) radar in the Antarctic to measure atmospheric phenomena. This project is called the Program of the Antarctic Syowa MST/IS radar (PANSY). This project uses approximately 1,000 antenna elements. The center frequency is 50 MHz and the diameter is approximately 160 m.

In the early 21st Century, in Japan, some trials as part of the development of high-efficiency phased arrays were carried out mainly for SPS applications. The Japanese SPS committee conducts development of the phased array. In FY2000, the SPS committee of the JAXA conducted a phased array experiment using solar cells for SPS called the solar power radio integrated transmitter (SPRITZ), which was developed mainly by Mitsubishi Heavy Industry (Figure 1.15) [MAT 02a]. Transmission was facilitated at a frequency of 5.77 GHz and the total number of elements was 100.

The other phased array field experiment was carried out by a team from Kobe University, Japan and John Mankins from Hawaii, USA in 2008. This group transmitted approximately 20 W of microwave power toward a target 150 km away using a phased array. Even though the microwave power in the experiment was not received, the transfer scheme has formed the basis of follow-up work.

A recent MPT project by Kyoto University's group in 2009 involved a field MPT experiment from an airship to the ground using two phased-controlled magnetrons (Figure 1.16) [MIT 10]. The experiment aimed to demonstrate an emergency wireless power source via MPT. The first phase-controlled magnetron was developed by Brown in the 1960s

[BRO 88]. The Kyoto University group revised the phase-controlled magnetron in the 1990s [SHI 00] and applied it to the airship experiment. Two magnetrons placed onboard the airship over 50 m above ground level were used as 2.45 GHz transmitters, and a mobile phone was charged with the wireless power.

Figure 1.15. *SPS demonstrator "SPRITZ" with 5.8 GHz transmission (demonstration in IAC2005)*

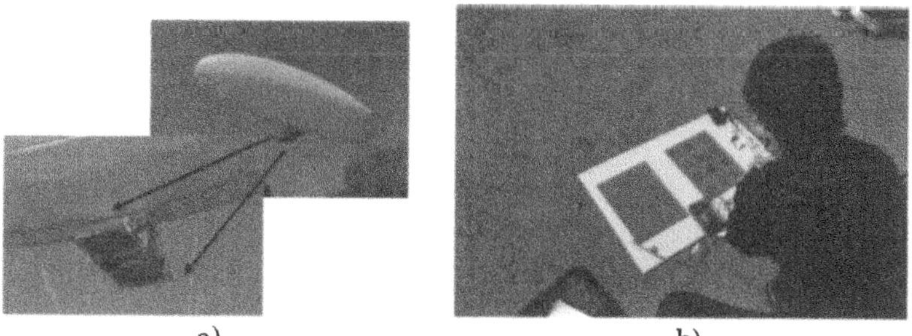

a) b)

Figure 1.16. *a) Emergency MPT experiment in Japan and b) wireless charging of a mobile phone on the ground [MIT 10]*

In addition to these MPT research projects, various kinds of MPT applications have been proposed, developed and commercialized worldwide. Details are described in Chapter 4.

The history of WPT is summarized in Figure 1.17. But there is little difference between the history of WPT and that of wireless communications. In fact, the history of WPT is largely based on the history of wireless communications.

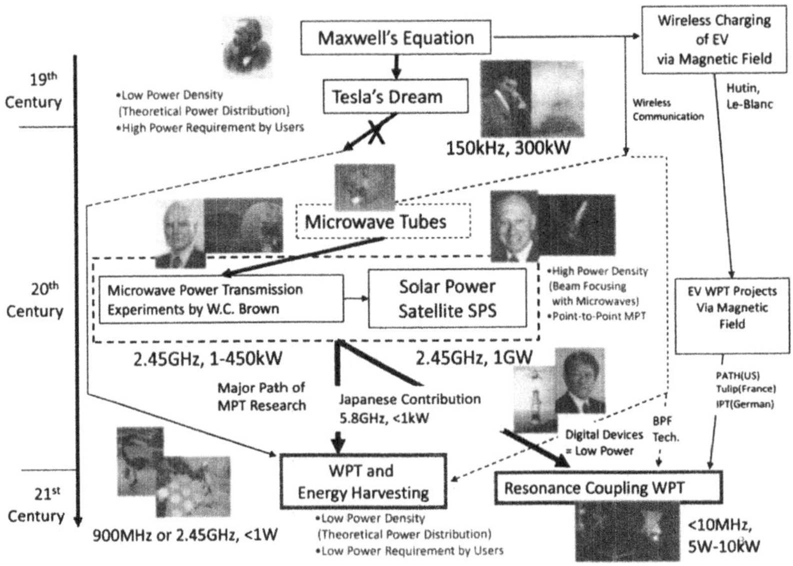

Figure 1.17. *Historical summary of WPT*

Chapter 2

Theory of WPT

2.1. Theoretical background

The following four Maxwell's equations can describe all electromagnetic phenomena such as radiowaves and light:

$$\nabla \times \mathbf{E} = -\frac{\partial \mathbf{B}}{\partial t} \text{ (Faraday's law of induction)} \quad [2.1]$$

$$\nabla \times \mathbf{H} = \mathbf{J} + \frac{\partial \mathbf{D}}{\partial t} \quad \text{(Ampère's circuital law)} \quad [2.2]$$

$$\nabla \cdot \mathbf{D} - \rho \text{(Gauss' law)} \quad [2.3]$$

$$\nabla \cdot \mathbf{B} = 0 \text{(Non-entity of Magnetic Charge)} \quad [2.4]$$

where $\mathbf{H}$, $\mathbf{B} = \mu\mathbf{H}$, $\mathbf{E}$, $\mathbf{D} = \varepsilon\mathbf{E}$, $\mathbf{J} = \sigma\mathbf{E}$ and ρ indicate magnetic field (A/m), magnetic flux density (T), electric field (V/m), electric flux density (C/m^2), current density (A/m^2) and charge density (C/m^3), respectively. μ, ε and σ indicate magnetic permeability (H/m), permeability (F/m) and electrical conductivity (1/Ω·m), respectively. Maxwell's equations indicate that electromagnetic waves propagate in a field based on relationships between the electric field, magnetic field, time and space.

It is known that electromagnetic energy is also associated with the propagation of electromagnetic waves. The energy of electromagnetic waves can be described as $\mathbf{E}\times\mathbf{H}$ (W/m^2). The vector $\mathbf{E}\times\mathbf{H}$ is called the Poynting vector. In plane wave propagation, the vector $\mathbf{E}\times\mathbf{H}$ results in a vector along the direction of propagation, which is same as the direction of the energy flow. It means that all electromagnetic waves themselves are energy. Theoretically, we can use all electromagnetic waves for wireless power transfer (WPT). Maxwell's equations indicate that the electromagnetic field and its power diffuse in all directions. Although we transmit energy in a communication system, the transmitted energy is diffused in all directions. WPT and wireless communication systems fundamentally differ only in how radiowaves are used by the receiver. In wireless communication systems, information is imparted to the transmitted radiowave via modulation. The radiowave is only a carrier of the information and is not very important itself. In contrast, for WPT systems, the radiowave itself is important and is used after rectifying. In wireless communication systems, the modulated radiowave is demodulated by a receiver. For WPT systems, the radiowave is rectified; in other words, it is converted from a high-frequency wave (greater than kilohertz) to a low-frequency wave (several tens of hertz) or DC.

2.2. Beam efficiency and coupling efficiency

2.2.1. *Beam efficiency of radiowaves*

An antenna is used to transmit and receive radiowaves. The antenna is a resonator at a selected frequency and is a converter of the radiowave from a circuit to free space (transmitting) and to a circuit from free space (receiving). The antenna is composed of metals and dielectric materials, whose shape and electric parameters decide the resonant frequency. Antennas are typically used for wireless communication and for radar and remote sensing. Identical antennas can also be used for WPT. Design theory and technologies are basically the same whether antennas are for wireless communication, radar, remote sensing or WPT.

The power of a radiowave is wirelessly transmitted from a transmitting antenna to a receiving antenna. It is a straightforward exercise to use the Friis transmission equation for calculating the receiving power at a far-field distance given as follows:

$$P_r = \frac{\lambda^2 G_r G_t}{(4\pi D)^2} = \frac{A_r A_t}{(\lambda D)^2} P_t \qquad [2.5]$$

where P_r, P_t, G_r, G_t, A_r, A_t, λ and D are the received power, transmitted power, antenna gain of the receiving antenna, antenna gain of the transmitting antenna, aperture area of the receiving antenna, aperture area of the transmitting antenna, wavelength, and the distance between the transmitting antenna and the receiving antenna, respectively (Figure 2.1). The Friis transmission equation assumes plane waves under sufficiently far-field conditions. The Friis transmission equation can be used to calculate the receiving power at far field for energy harvesting or for WPT using diffused radiowaves.

However, we cannot use the Friis transmission equation for a WPT application by using beamed radiowaves because we must calculate the receiving power at a distance in the near field. A spherical wave is generated in the near field where beamed WPT is applied.

Therefore, we use the following τ parameter to calculate the receiving power or beam efficiency (BE) η [GOU 61, BRO 73a, ITU 00].

$$\tau^2 = \frac{A_t A_r}{(\lambda D)^2} \quad [2.6]$$

$$\eta = \frac{P_r}{P_t} = 1 - e^{-\tau^2} \quad [2.7]$$

It is assumed that the amplitude and the phase of a radiowave are uniform at the transmitting antenna, as shown in Figure 2.1. It is also assumed that the transmitting antenna and the receiving antenna are at the correct position.

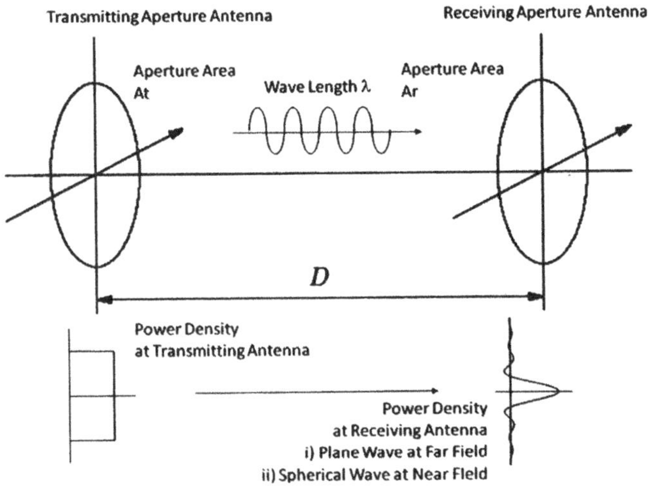

Figure 2.1. *Transmitting antenna and receiving antenna*

From equation [2.6], it can be observed that τ^2 represents the BE calculated by the Friis transmission equation [2.5] in the far field. Equation [2.7] indicates that, for a given frequency and distance, a large aperture antenna must be used to increase the BE. The BE in the far field and the near field using the τ parameter is shown in Figure 2.2. We can increase the BE approximately to 100% with $\tau > 2$ in the near field. The theory does not depend on power. Therefore, we can transmit high power via radiowaves. For small τ, the

BE obtained is almost the same from both formulas, and under those conditions, we can assume a plane wave in the far field. When the distance between the transmitting antenna and the receiving antenna is shortened, τ becomes larger and we cannot assume a plane wave at the receiving antenna plane, and therefore we cannot use the Friis transmission equation.

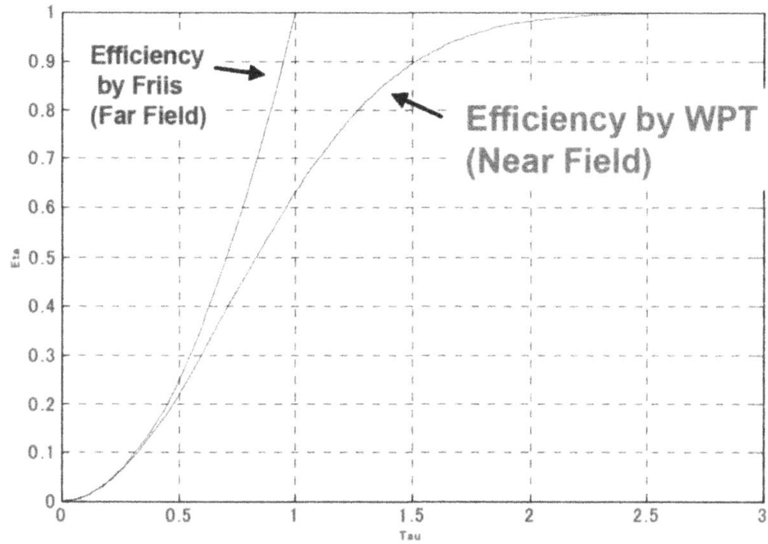

Figure 2.2. *Beam efficiency in the far field and the near field using the τ parameter*

Equation [2.7] is an approximation that neglects sidelobes. We can calculate the BE exactly by including the near-field beam pattern. Figure 2.3 shows the BE of a WPT system, e.g. the wireless charging of an electric vehicle, where f, Tx and Rx are frequency, the diameter of the transmitting antenna and the diameter of the receiving antenna, respectively. The BE is 90% at a distance of <5 m for standard WPT parameters. Therefore, the efficiency of WPT via radiowaves is sufficiently high relative to resonant or inductive coupling.

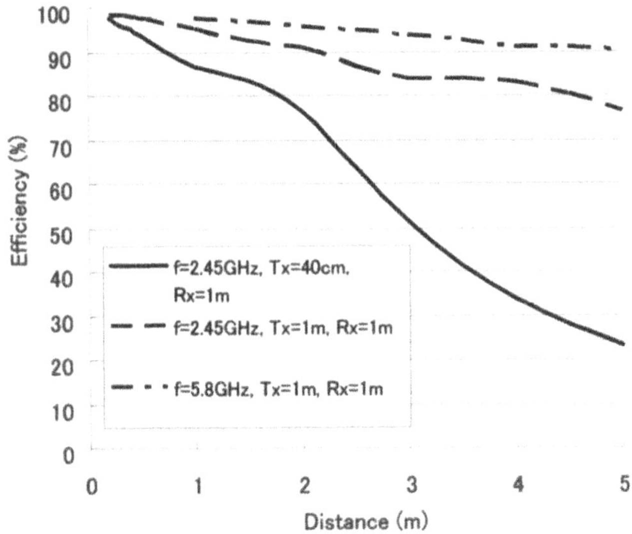

Figure 2.3. *Beam efficiency of WPT system without approximation*

The use of microwaves is essential to realize $\tau > 2$ because τ is inversely proportional to the wavelength. The first WPT experiment conducted by Nicola Tesla in the early 20th Century had a drawback of low BE because he used 150 kHz radiowaves. In the 1960s, Brown's WPT experiments succeeded because he used microwaves in the gigahertz range having a high BE as the wavelength of microwaves is much shorter than that of 150 kHz radiowaves. However, DC–RF conversion and RF–DC conversion efficiencies decrease at higher frequencies. Therefore, the total efficiency of WPT via microwaves is approximately 50%, which includes DC–RF conversion efficiency, BE, absorption efficiency of the antenna and the RF–DC conversion efficiency.

2.2.2. *Theoretical increase of beam efficiency*

From a theoretical perspective, we cannot increase the BE in the manner indicated by equation [2.7], but we can increase the BE by changing the power distribution on the transmitting antenna. We assume that the power distribution on the transmitting antenna is uniform in equation [2.7], as shown in Figure 2.1. A well-known power distribution exists that will enable us to increase the BE, which can be described with the following formula.

1) Gaussian power distribution

$$a(n) = \exp\left[-(\ln R)\left\{ 2\frac{n-1}{N-1} - 1 \right\}^2 \right] \qquad [2.8]$$

where $a(n)$ is the power distribution on the transmitting antenna, R is the ratio between the power at the center and at the edge of the transmitting antenna, n is the index of an antenna element and N is total number of antenna elements. In equation [2.8], we assume that an aperture antenna can be considered as a sum of antenna elements of a sufficient number N, which is a description of a phased array. It is standard to call a distribution "10 dB Gaussian power distribution" when the value of R is 10 dB. The beam pattern created by the Gaussian power distribution has a lower gain and wider beam width than that created by a uniform power distribution. As a result, the BE determined by the Gaussian power distribution increases relative to that by the uniform power distribution. The beam patterns of Gaussian power distributions and uniform power distributions at far field are shown in Figure 2.4. The Gaussian power distribution is often adopted in a solar power satellite (SPS) design to increase the BE.

2) Chebyshev power distribution

$$a_n = \begin{cases} \dfrac{2\zeta}{(N-1)\sqrt{1-\xi^2}} I_1(\zeta\sqrt{1-\xi^2}) & (\zeta \neq 1) \\ 1 \quad (\zeta = 1) & \end{cases} \qquad [2.9]$$

where R is the ratio between the power at the center and the power at the edge of the transmitting antenna, n is the index of an antenna element, N is total number of the antenna elements and

$$\xi = \frac{2n - N - 1}{N - 1} \quad [2.10]$$

$$\zeta = (N-1)\tanh\left[\frac{\ln(R - \sqrt{R^2 - 1})}{N - 1}\right] \quad [2.11]$$

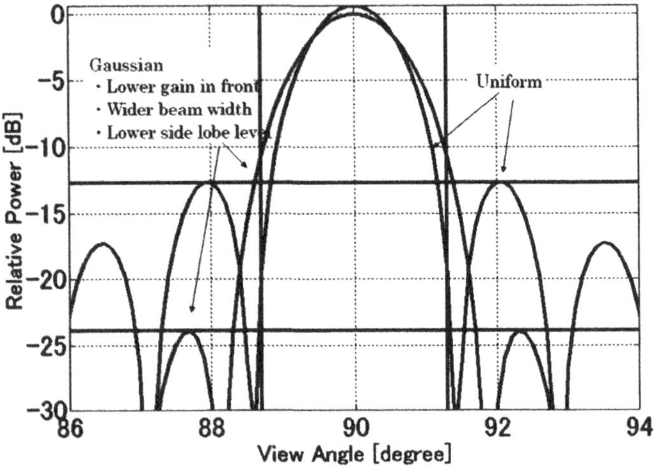

Figure 2.4. *Comparison of beam patterns at far field between a Gaussian power distribution and a uniform power distribution*

The beam pattern created by a Chebyshev power distribution is formulated as follows:

$$C(\theta) = \cos\{(N-1)\cos^{-1}(z\cos\frac{u}{2})\} \quad [2.12]$$

where d_n is the element spacing and

$$u = 2\pi\frac{d_n}{\lambda}\sin\theta \quad [2.13]$$

$$z = \cosh\{\frac{1}{N-1}\ln(R + \sqrt{R^2 - 1})\} \quad [2.14]$$

Figure 2.5 illustrates a beam pattern at far field created by the Chebyshev power distribution with $N = 50$ and $R = 25$ dB. It is a characteristic of the beam pattern generated by the Chebyshev power distribution that all sidelobe levels are the same. The first sidelobe level is suppressed, but the other sidelobe levels are increased.

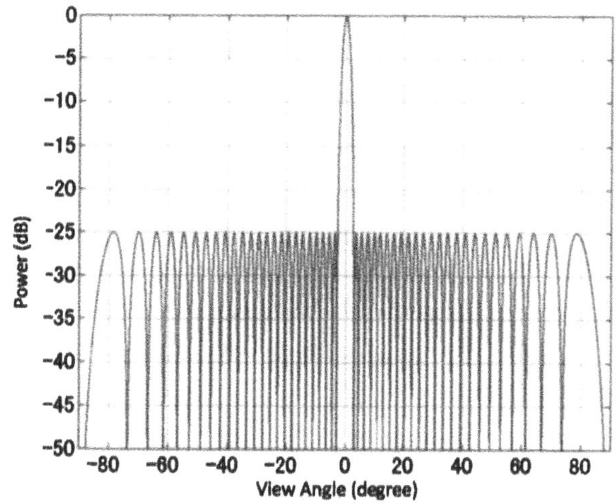

Figure 2.5. *Beam pattern at far field created by the Chebyshev power distribution with N = 50 and R = 25 dB*

3) Taylor power distribution

$$a(u) = \frac{1}{2\pi}\int_{-\infty}^{\infty} T(u)e^{-juv}du \qquad [2.15]$$

where

$$u = 2\pi\frac{d_n}{\lambda}\sin\theta \qquad [2.16]$$

$$T(u) = g(u)\prod_{m=1}^{\bar{n}}\left[\frac{1-\left(\dfrac{u}{u_{cm}}\right)^2}{1-\left(\dfrac{u}{u_{gm}}\right)^2}\right] \qquad [2.17]$$

where

$g(u)$: directivity from the uniform power distribution;

u_{cm}: mth null point from the Chebyshev power distribution;

u_{gm}: mth null point from the uniform power distribution;

$\bar{n}$: same number of null points derived from the Chebyshev power distribution.

The Taylor power distribution is a combination of the Chebyshev and the uniform power distributions. Sidelobe levels that are closer to a main lobe than the mth number are similar to that from the Chebyshev power distribution. A beam pattern at far field created by the Taylor power distribution with $N = 50$, sidelobe levels of –25 dB and $\bar{n} = 1$ is shown in Figure 2.6.

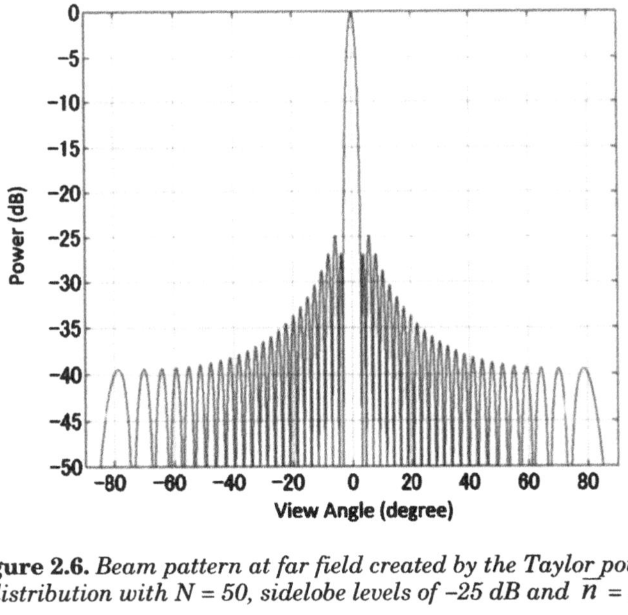

Figure 2.6. *Beam pattern at far field created by the Taylor power distribution with $N = 50$, sidelobe levels of –25 dB and $\bar{n} = 1$*

2.2.3. *Coupling efficiency at very close coupling distance*

We usually use antennas at near field or far field for WPT via radiowaves. High frequencies in the megahertz to gigahertz range are beneficial to effectively transmit wireless power. If we use lower frequencies in the range of less than a kilohertz, it is difficult to make use of the WPT system even at near field. A distance closer than near field is referred to as the reactive field. At the reactive field, we should consider not an electromagnetic wave but an electric field or a magnetic field alone. When we use a coil as a closed loop, the coil creates a magnetic field and the magnetic field transfers wireless power. This is inductive coupling WPT. Inductive coupling WPT is based on Ampère's circuital law and Faraday's law of induction. Ampère's law and Faraday's law of induction are approximations of Maxwell's equations. Ampère's circuital law describes the relationship between the integrated magnetic field around a closed loop (coil) and the electric current passing through the loop. Faraday's law of induction describes the relationship between a time-varying magnetic field and an induced electric field. The electric power is carried through the magnetic field between two coils. Ampère's circuital law and Faraday's law of induction are both examples of Maxwell's equations. The efficiency of WPT depends on the coupling coefficient, which in turn depends on the distance between the two coils. Therefore, wireless energy cannot be carried over a distance longer than a few millimeters with high efficiency using inductive coupling, and the frequency used in inductive coupling is below a few dozen megahertz.

When we add a capacitor of a capacitance (C) to a coil that is defined by an inductance (L), the two elements form a resonator having properties defined by both C and L. When two resonators are electromagnetically coupled, the energy in one resonator is transmitted to the other through an evanescent mode wave. This phenomenon is well known as coupling theory applied to a microwave band pass filter.

However, it was not until 2006 that Massachusetts Institute of Technology (MIT) researchers demonstrated a WPT experiment using resonant coupling. Resonant coupling with coils is called magnetic resonant coupling. The transmitted power is mainly a magnetic field supported by a coil. Resonant coupling is realized with two plane-shaped conductors passing through an electric field and is called electric resonant coupling.

A transmitting antenna and a receiving antenna are usually not coupled, and wireless power is transferred between them through space. Two coils or two resonators, however, are electromagnetically coupled for inductive coupling WPT and resonant coupling WPT systems. A theoretical investigation based mainly on inductive coupling has been conducted for magnetic resonant coupling. From the theory of inductive coupling, $k{\cdot}Q$, where k is the coupling coefficient and Q is the quality factor of the coil resonator, is the critical factor established. The maximum coupling transmission efficiency η is calculated using $k{\cdot}Q$, as shown in the following equations.

$$\eta = \frac{form^2}{\left(1+\sqrt{1+form^2}\right)^2}$$
$$form^2 = k^2 Q_1 Q_2 = \frac{\omega^2 M^2}{R_1 R_2} \qquad [2.18]$$

where ω is the frequency, R is the resistance and M is the mutual inductance. The efficiency curve is shown in Figure 2.7. The coupling transmission efficiency is determined by $k{\cdot}Q$. In inductive coupling, a high Q factor cannot be used; therefore, k should increase as a function of the distance between the two coils. However, in resonant coupling, it is easy to increase $k{\cdot}Q$ with a high Q factor even if the distance between the two coils is large and the k factor is small (note that k contains a wavelength parameter). If a lower frequency is selected, then k increases if the distance and Q factor are the same. As a result, the WPT distance having a high efficiency is expanded using a lower frequency in an inductive or resonant coupling WPT system. A lower frequency indicates that higher efficiency can be achieved using lower cost components for the system.

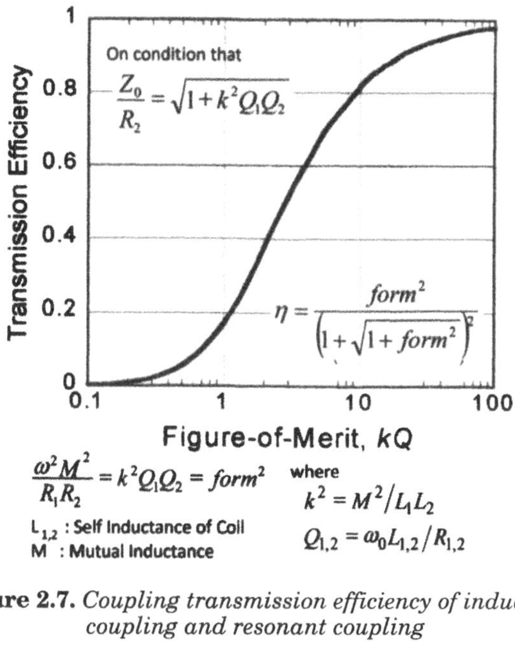

$$\frac{\omega^2 M^2}{R_1 R_2} = k^2 Q_1 Q_2 = form^2 \quad \text{where} \quad k^2 = M^2/L_1 L_2$$

$L_{1,2}$: Self Inductance of Coil

M : Mutual Inductance

$$Q_{1,2} = \omega_0 L_{1,2}/R_{1,2}$$

Figure 2.7. *Coupling transmission efficiency of inductive coupling and resonant coupling*

2.3. Beam forming

2.3.1. *Beam-forming theory for the phased array and its error*

Equation [2.18] includes a parameter reflecting the positions of the resonators or coils in k. The value of k includes not only the positions, but also the distance between the two resonators or two coils. Equation [2.6] or [2.7] includes a parameter reflecting the distance between two antennas but does not include the positions of the two antennas. The easiest way to maintain a high BE is to control the direction of the antenna and maintain the correct position. However, we have to mechanically control the direction of the antenna, which is not an easy task.

A phased array antenna is a useful piece of technology for electrically controlling the beam direction. The phased array is composed of plural antennas as shown in Figure 2.8. We control the phase and the amplitude of the radiowave transmitted at each antenna with phase shifters or a beam-forming network circuit. A beam form is controlled by interference of the radiowaves (Figure 2.9). To calculate the beam pattern from a phased array, we need r (the distance between a transmitting antenna element and a receiving point), θ and ϕ (the two angles acting between a transmitting antenna element and a receiving point) and $D(r,\theta,\phi)$ (the element pattern of the transmitting antenna element). All antenna elements have a peculiar element pattern (directivity) $D(r,\theta,\phi)$.

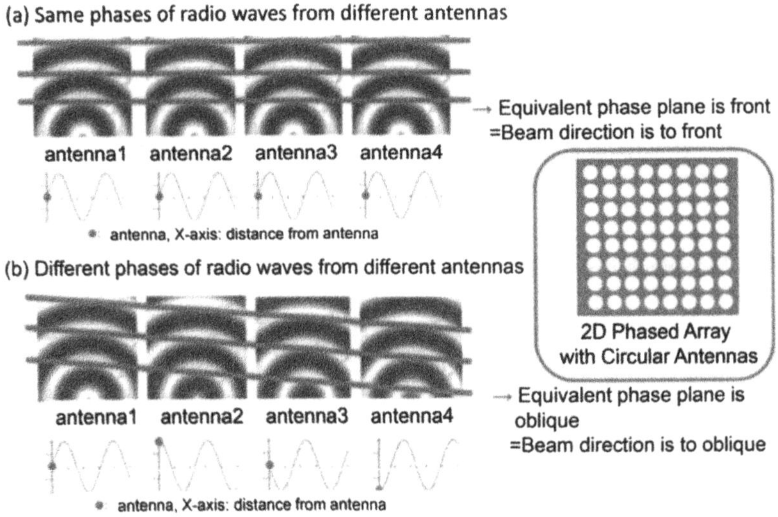

Figure 2.8. *Concept of a phased array I*

The calculation method of the phased array shown in Figure 2.9 can be used both in near field and far field. However, it is difficult to calculate the beam form. Therefore, at far field, we can omit the distance parameter r and use the following equation to calculate the beam form $E(\theta,\phi)$. It is based on consideration of the "array factor". The array factor $A(\theta,\phi)$ collectively indicates the directivity of all antenna elements, which considers the phased array as an aperture array. We can calculate the beam form $E(\theta,\phi)$ as the product of the element factor $D(\theta,\phi)$ and the array factor $A(\theta,\phi)$ at far field.

$$E(\theta,\phi) = D(\theta,\phi)A(\theta,\phi) \quad [2.19]$$

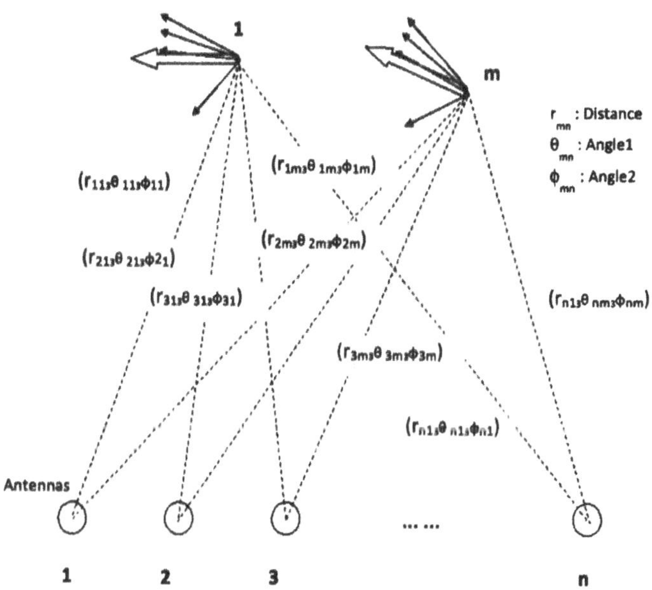

Figure 2.9. *Concept of a phased array II*

The array factor is determined by the antenna elements by their positions, amplitude and phase, and is not related to the types of antennas used. When we consider a one-dimensional (1D) uniformly spaced array of N antenna elements, the array factor is given as follows:

$$A(\theta,\phi)=\sum_{n=1}^{N}a_n e^{j\varphi_n} \quad [2.20]$$

where a_n and ϕ_n are the amplitude and the phase of nth antenna element, respectively. The parameters are shown in Figure 2.10. Equation [2.19] is transformed using equation [2.19] to obtain the following equation:

$$E(\theta,\phi)=D(\theta,\phi)A(\theta,\phi)=D(\theta,\phi)\cdot\sum_{n=1}^{N}a_n e^{j\varphi_n} \quad [2.21]$$

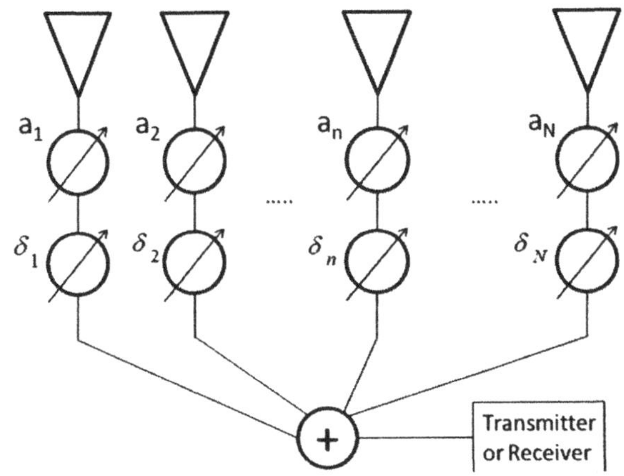

Figure 2.10. *Parameters of array used to calculate the beam form at far field*

The phase of an antenna element is caused by the difference of the antenna position and a phase shifter that can arbitrarily control the phase. The phase of the nth element can be geometrically described as $\phi_n = kd_n\cos\theta+\delta_n$, where k is the wave number, d_n is the nth element spacing and δ_n is an additional phase shift, for example, caused by a phase shifter. Equation [2.21] can be described using ϕ_n as follows:

$$E(\theta,\phi) = D(\theta,\phi)\,A(\theta,\phi) = D(\theta,\phi)\cdot\sum_{n=1}^{N} a_n e^{j(kd_n\cos\theta+\delta_n)} \qquad [2.22]$$

The parameters describing a phased array are shown in Figure 2.11. The relationships between the beam form of a phased array, element factor and array factor are shown in Figure 2.12. It is easy to calculate the beam form at far field using equation [2.22].

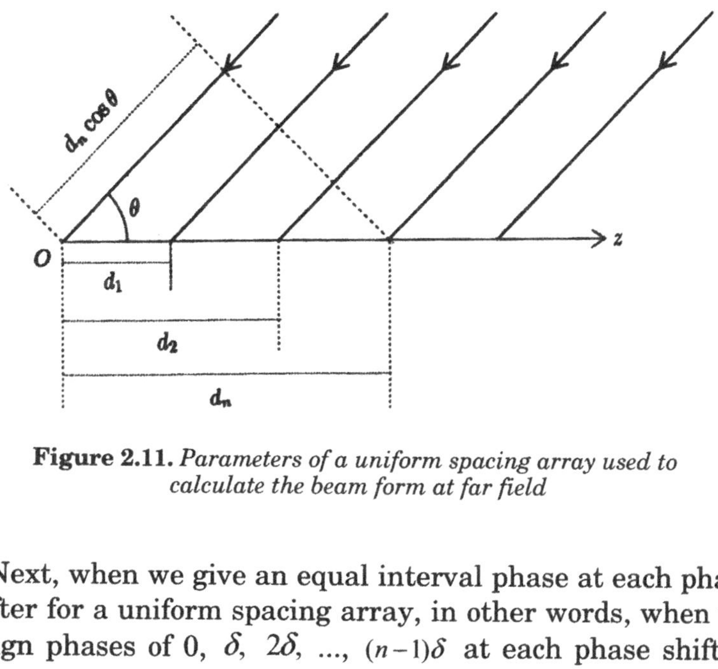

Figure 2.11. *Parameters of a uniform spacing array used to calculate the beam form at far field*

Next, when we give an equal interval phase at each phase shifter for a uniform spacing array, in other words, when we assign phases of 0, δ, 2δ, ..., $(n-1)\delta$ at each phase shifter, the array factor $A(\Psi)$ is given as follows:

$$A(\psi) = \frac{\sin(\frac{N\psi}{2})}{N\sin(\frac{\psi}{2})} \qquad [2.23]$$

where a_n is equal to 1 and Ψ is an angle. The beam form using $A(\Psi)$ is shown in Figure 2.13 when a phase standard point is at the center of the array and the amplitude is standardized by a maximum corresponding to a magnitude of 1. The figure corresponds to a beam form from an aperture antenna with a uniform power distribution. Sidelobes are also shown.

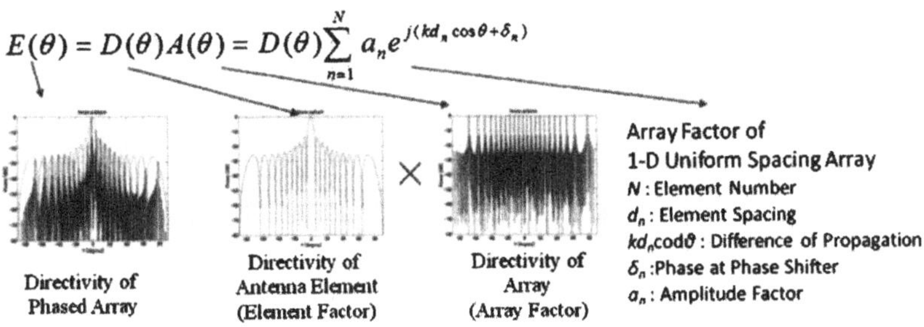

Figure 2.12. *Relationships between the beam form of a phased array, element factor and array factor*

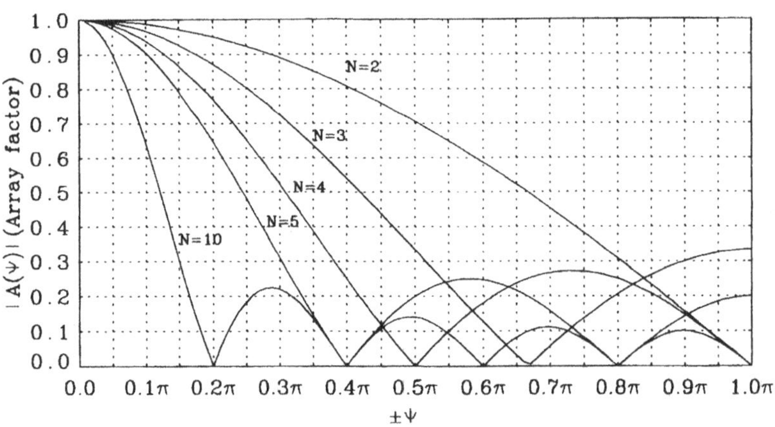

Figure 2.13. *Beam form of uniform spacing array*

The direction of maximum beam power θ_{max} is given by equation [2.22] as follows:

$$\Psi = kd\cos\theta + \delta = 0 \qquad [2.24]$$

Equation [2.24] contains a sinusoidal function. It is easy to understand that a lot of solutions are realized from equation [2.24]. This equation indicates that a number of intense beam points will occur when an equal interval phase is given at each phase shifter in a uniform spacing array. This is called a grating lobe, whose power is the same as the main beam. Figures 2.14(a) and (b) show the two beam forms of a 1D uniform spacing array. N_a corresponds to an element spacing d. A large Na corresponds to a large d. Figure 2.14(a) shows a beam form where the phase at all phase shifters is 0. The beam is focused at the center of the array. Figure 2.14(b) shows a beam form wherein we assign an equal interval phase at each phase shifter to focus the beam at an angle of 5°. In Figure 2.14(b), we can see that grating lobes arise. Power is distributed to all grating lobes when grating lobes arise. It is clear that grating lobes should be suppressed in order to realize high BE for the WPT system.

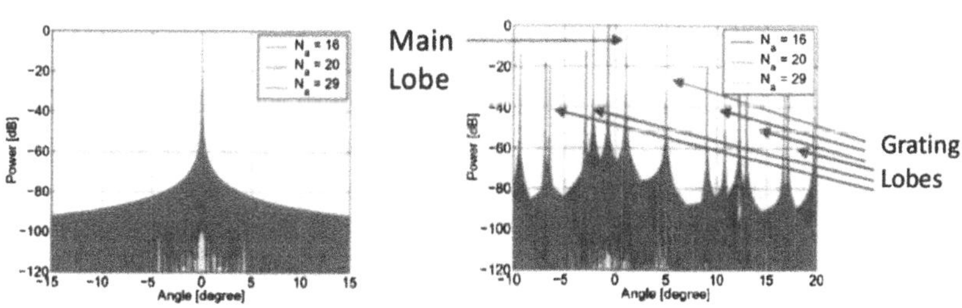

a) b)

Figure 2.14. *Beam form of a 1D phased array: a) focus direction: 0°, b) focus direction: 5°. For a color version of this figure, see www.iste.co.uk/shinohara/radiowaves.zip*

The condition of no grating lobes is given by the following equation:

$$d < \frac{\lambda}{1+\sin|\theta_s|} \qquad [2.25]$$

where θ_s is the focused direction angle, d is the element spacing and λ is the wavelength. For example, $\theta_s < 19.5^\circ$ when $d=0.75\lambda$ and $\theta_s < 41.8^\circ$ when $d=0.6\lambda$. No grating lobes arise in the condition $d<0.5\lambda$. If $d>\lambda$, grating lobes must rise except in the case $\theta_s = 0^\circ$, i.e. for a front beam.

Phased array and beam-forming theories do not account for phase/frequency/amplitude error. The efficiency of the array must decrease owing to phase/frequency/amplitude error. Phase/frequency/amplitude error for a phased array causes differences in beam direction and the rise of sidelobes. Differences in the beam direction caused by phase errors are described by the following equation:

$$\Delta B = \frac{12}{N^3}\Delta\phi^2 \qquad [2.26]$$

where ΔB is the change in the beam direction, N is number of phased array elements and $\Delta\phi$ is the maximum antenna element phase error. For a sufficiently large number of elements, the change in the beam direction is negligible. For example, assuming N = 10,000 in one dimension and $\Delta\phi = 5^\circ$ in the SPS case, ΔB is $1.732 \times 10^{-5\circ}$ or 10.9 m at a distance of 36,000 km.

A more serious problem for the decline in the beam collection efficiency is the rise of sidelobes caused by phase/frequency/amplitude error. The following equation defines the mean squared sidelobe level (MSSL) by means of the probability of antenna element trouble (P_T), dispersion of the amplitude error ($\sigma_A{}^2$), dispersion of the phase error ($\sigma_\phi{}^2$), aperture efficiency (η_a) and the number of elements (N) [SKO 90].

$$MSSL = \frac{(1 - P_T) + \sigma_A^2 + P_T \sigma_\varphi^2}{\eta_a P_T N} \qquad [2.27]$$

The MSSL indicates an average error sidelobe level caused by antenna element trouble and errors in amplitude and phase. It is assumed that the MSSL results in power leakage from the main lobe and is the average sidelobe level in all directions. The MSSL must be added on an original sidelobe level without any errors in order to know the real sidelobe level with the errors. The MSSL gives a probability (P) that there is no sidelobe over a value R_T as shown in the following equation:

$$\begin{aligned} &P[\text{N sidelobes} < R_T] = \{1\text{-}P[\text{one sidelobe} > R_T]\}^N \\ &\cong 1 - N \times P[\text{one sidelobe} > RT] \end{aligned} \qquad [2.28]$$

For example, when the MSSL is –47 dB, the probability is 0.99 that sidelobes do not rise above –40 dB (Figure 2.15).

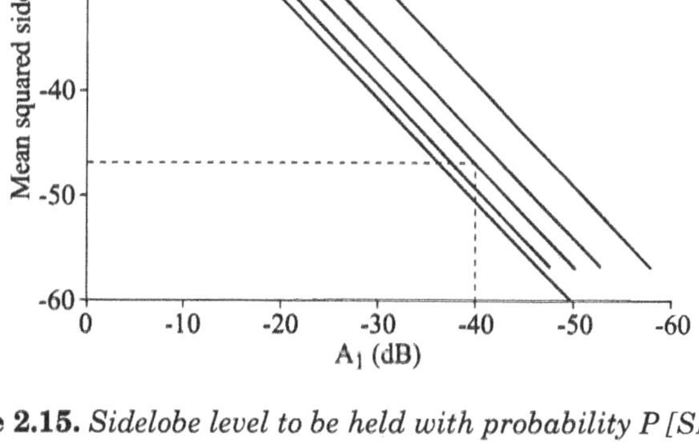

Figure 2.15. *Sidelobe level to be held with probability P [SKO 90]*

The rise of sidelobes decreases the antenna gain and beam collection efficiency. The antenna gain (G) with probability of antenna element trouble (P_T), dispersion of amplitude error (σ_A^2) and dispersion of phase error (σ_ϕ^2) can be described as follows [MAI 94]:

$$G = G_0 \frac{P_T}{1 + \sigma_A^2 + \sigma_\phi^2} \qquad [2.29]$$

where G_0 is the antenna gain without any errors.

If antenna planes structurally separate from each other, grating lobes, whose power levels are the same as the main beam, may occur and power may not be effectively concentrated to the rectenna array. This problem occurs in module-type phased arrays. The idea of a random array has been investigated as a means of suppressing grating lobes. However, as the sidelobe level increases, beam collection efficiency decreases and then special techniques are required. The power of grating lobes diffuses not to the main lobe but to the sidelobes. Therefore, we have to fundamentally suppress grating lobes for an efficient microwave power transfer (MPT) system.

2.3.2. *Target detecting via radiowaves*

Control of the direction of a power beam is of no significance if we cannot detect the receiving target. There are various target detection methods: global positioning system (GPS); optical detection; direction-of-arrival (DOA) methods, such as multiple signal classification (MUSIC) via radiowaves; and a method named retrodirective detection, which uses a pilot signal radiowave. In the microwave lifted airplane experiment (MILAX) experiment shown in Figure 1.7, optical target detection was accomplished using two charge coupled device (CCD) cameras. The difference between the retrodirective method and other target detection methods is that only the former can detect both the positions of the target and antenna elements (the shape of the antenna array) in a phased array.

Use of radiowaves as pilot signals to detect the receiving target has been made in various MPT field experiments. In certain MPT experiments, a DOA algorithm with a pilot signal and phase shifters controlled by the DOA algorithm were used to implement a phased array. The minimum configuration of a DOA radiowave array is a two-antenna system, as shown in Figure 2.16. The DOA angle θ can be estimated from the phase difference between the two antennas $\Delta\theta$, element spacing d and wavelength λ as follows:

$$\theta = \sin^{-1}\left(\frac{\lambda \cdot \Delta\theta}{2\pi d}\right) \quad (rad) \qquad [2.30]$$

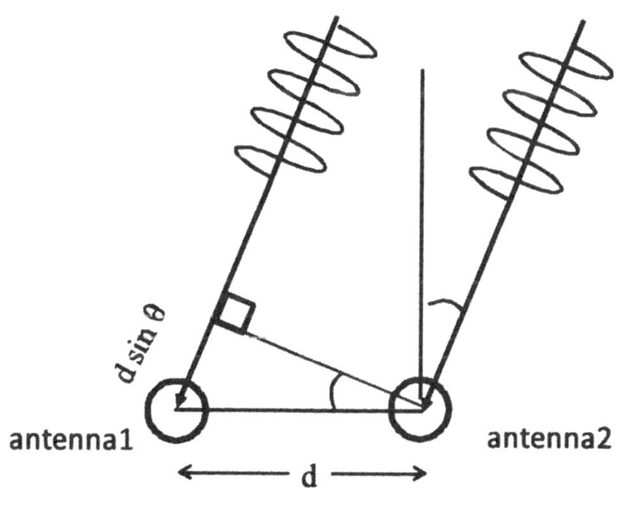

Figure 2.16. *Fundamental direction of arrival of radiowave*

After the estimation of the DOA angle, each phase of the antennas is estimated. A power beam is created with the phased data and is sent back to the detected source. Phase shifters must be used to control the beam produced by each antenna in typical phased arrays. For optimum beam forming in a phased array, various methods such as neural networks, genetic algorithms and multiobjective optimization learning are available. The term "optimum" has multiple meanings: to suppress sidelobe levels, to increase beam collection efficiency and to produce multiple power beams. Array optimization can be freely implemented by considering system complexity and the calculation time required by the control algorithm. Such DOA systems that use control algorithms are sometimes called "software retrodirective" systems.

However, an error occurs in DOA estimation because of the signal/noise (SN) ratio. The estimation error $\Delta\theta'$ is formulated as follows [LIP 87]:

$$\Delta\theta' = \frac{\lambda}{\pi d \cos\theta \sqrt{SN}} \quad (rad) \qquad [2.31]$$

DOA estimation errors calculated by equation [2.35] for various SN ratios are shown in Figure 2.17. DOA estimation errors should be considered in the design of an MPT system. If more than two antennas are used, SN ratio of the system decreases. An advanced algorithm, such as MUSIC, can also be applied to implement the DOA. The algorithm can decrease SN ratio, thus increasing the target detection accuracy.

On the other hand, phase shifters are not used in a retrodirective system. A pilot signal originating from the target is used in both retrodirective and software retrodirective systems. A corner reflector is the most basic retrodirective system [SUN 03]. Corner reflectors consist of perpendicular metal sheets jointed at the apex (Figure 2.18(a)). Incoming signals are reflected back in the DOA through multiple reflections off the reflector wall. The Van Atta array is another basic retrodirective system

[VAN 59]. It is made up of pairs of antennas equidistantly spaced from the center of the array and connected with equilength transmission lines (Figure 2.18(b)). The signal received by an antenna is re-radiated by its pair; thus, the order of re-radiating elements is inverted with respect to the center of the array, achieving proper phasing for retrodirectivity.

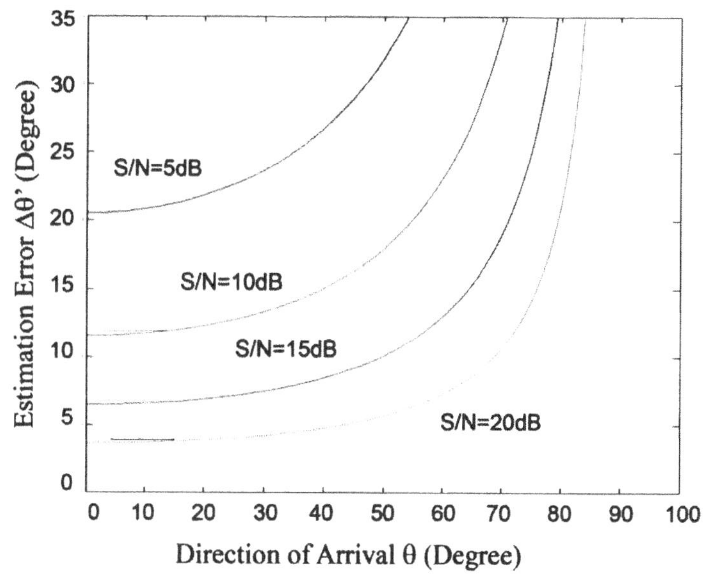

Figure 2.17. *Estimation error of DOA Δθ*

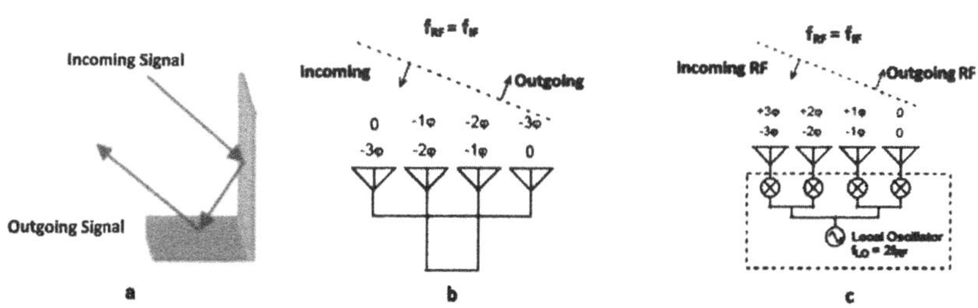

Figure 2.18. *a) Two-sided corner reflector, b) Van Atta array, c) retrodirective array with phase-conjugate circuits [SUN 03]*

Usually, retrodirective systems have phase-conjugate circuits in each receiving/transmitting antenna (Figure 2.18(c)), which play the same role as the pairs of antennas spaced equidistantly from the center of the array in a Van Atta array. A pilot signal transmitted from the target (amplitude: V_{RF}, angular frequency: ω_{RF} and phase at each antenna: $+\theta_n$) is received and re-radiated through the phase-conjugate circuit in the direction of the target. A local signal (amplitude: V_{LO} and angular frequency: ω_{LO}) is used for phase conjugation. Two forms of the pilot signal and the local signal are mixed and the conjugate signal V_{IF} is obtained as defined by the following equation and as shown in Figure 2.19(a):

$$\begin{aligned} V_{IF} &= V_{RF}\cos\left(\omega_{RF}t+\theta_n\right)\cdot V_{LO}\cos\left(\omega_{LO}t\right) \\ &= \frac{1}{2}V_{RF}V_{LO}\left(\cos\left[\left\{\omega_{LO}-\omega_{RF}\right\}t-\theta_n\right]+\cos\left[\left\{\omega_{LO}+\omega_{RF}\right\}t+\theta_n\right]\right) \end{aligned} \qquad [2.32]$$

When a low-pass filter is used after mixing, the following signal is obtained. The phase $+\theta_n$ is changed to $-\theta_n$:

$$V_{IF} = \frac{1}{2}V_{RF}V_{LO}\cos\left[\left\{\omega_{LO}-\omega_{RF}\right\}t-\theta_n\right] \qquad [2.33]$$

If we choose $\omega_{LO} = 2\omega_{RF}$, equation [2.33] becomes:

$$V_{IF} = \frac{1}{2}V_{RF}V_{LO}\cos\left(\omega_{RF}t-\theta_n\right) \qquad [2.34]$$

as shown in Figure 2.19(b).

The accuracy of the sending direction depends on the stability of the frequency of the pilot signal and the local oscillator (LO) signal. To decrease a beam-forming error caused by the fluctuation of the LO signal, the same pilot signal and frequency doubler are used instead of the LO signal.

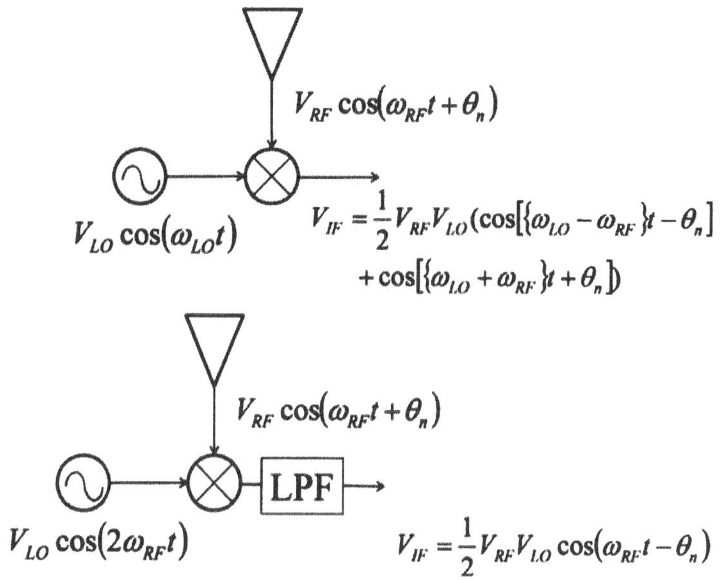

Figure 2.19. *Block diagram of a retrodirective system with phase conjugation: a) output without filtering; b) output with low-pass filter (LPF)*

2.4. Beam receiving

Equation [2.3] represents only the transfer efficiency. Therefore, we have to additionally consider the radiation or absorption efficiency of an antenna for WPT via radiowaves.

An infinite array can theoretically absorb 100% of a transmitted radiowave [DIA 68, STA 74]. In general, the infinite array model can approximately apply to the analysis of a large array. Although edge effects cannot be evaluated under such an approximation, very simple and useful results are obtained.

Itoh *et al.* calculated the absorption efficiency of a rectenna array composed of circular microstrip antennas (CMSA), as shown in Figure 2.20. The following analysis is referenced in [ITO 86]. An angle θ representing the direction of propagation of the incident plane wave is restricted to the x–z plane. Polarization is parallel to the x–z plane. Therefore, the polarization mismatch loss can be ignored. Here, it is assumed that the thickness of the substrate is very small as compared to the wave length. Therefore, a single-mode (TM_{110}) excitation model of a magnetic current loop on the ground plane, a very simplified model, can be considered. Using Stark's method [STA 74], the active admittance Y is obtained as follows:

$$\begin{aligned} Y = \sum_m \sum_n \Bigg[& \frac{\pi^2 a^2}{Z_0 L_x L_y} \cdot \frac{k^2 - h_n^2}{k\gamma_{mn}^*} \cdot \left\{ J_0(\rho) + J_2(\rho) \cdot \cos 2\alpha \right\}^2 \\ & - \frac{2\beta_m h_n}{k\gamma_{mn}^*} \cdot \left\{ J_2(\rho) \cdot \sin 2\alpha \right\}^2 \cdot \left\{ J_0(\rho) + J_2(\rho) \cdot \cos 2\alpha \right\} \\ & + \frac{k^2 - \beta_m^2}{k\gamma_{mn}^*} \cdot \left\{ J_2(\rho) \cdot \sin 2\alpha \right\}^2 \Bigg] \end{aligned} \quad [2.35]$$

where $\beta_m = 2\pi m / L_x + k \sin\theta$, $h_n = 2\pi n / L_y$,

$$\gamma_{mn} = \sqrt{k^2 - \beta_m^2 - h_n^2} \quad \left(\mathrm{Re}\{\gamma_{mn}\} > 0, \mathrm{Im}\{\gamma_{mn}\} < 0 \right),$$

$$\rho = a\sqrt{\beta_m^2 + h_n^2}\ , \ \alpha = \tan^{-1}(\beta_m / h_n)$$

k is the wave number of the indecent wave, and β_m, h_n and γ_{mn} are propagation constants of the space harmonic wave along x, y and z, respectively. Here higher waves correspond to grating lobes.

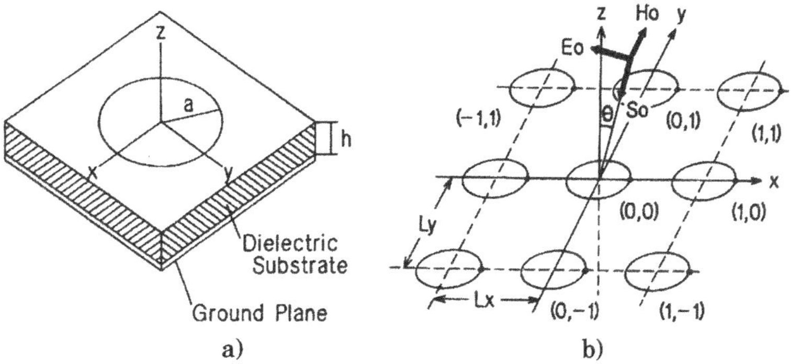

Figure 2.20. *a) Circular microstrip antenna; b) geometry of infinite CMSA array [ITO 86]*

When all higher waves are evanescent, the active conductance G is expressed using equation [2.35] as follows:

$$G = \frac{\pi^2 a^2}{Z_0 L_x L_y \cos\theta} \cdot \{J_0(\rho) - J_2(\rho)\}^2 \qquad [2.36]$$

where $\rho = ka\sin\theta$.

The absorption efficiency of the infinite rectenna array is defined as the ratio of the maximum receiving power of an element to the incident power per element [ADA 81]. Then, the absorption efficiency η_{absorp} is represented by:

$$\eta_{absorp} = \frac{A_e}{L_x L_y \cos\theta} \qquad [2.37]$$

Here, the numerator denotes the absorption cross-section. The denominator denotes the cross-section of the cell. The absorption cross-section A_e is given by:

$$A_e = \frac{G_{rad}}{G} \cdot \frac{\lambda^2}{4\pi} D_a \qquad [2.38]$$

where G_{rad} and D_a are the radiation conductance and the directivity of a single element, respectively, and G is the active conductance. Therefore, A_e is rearranged as follows:

$$A_e = \frac{\pi^2 a^2}{Z_0 G} \cdot \{J_0(\rho) - J_2(\rho)\}^2 \qquad [2.39]$$

where $\rho = ka \sin\theta$.

From equations [2.37] and [2.39], the absorption efficiency is obtained as follows:

$$\eta_{absorp} = \frac{\pi^2 a^2}{Z_0 G L_x L_y \cos\theta} \cdot \{J_0(\rho) - J_2(\rho)\}^2 \qquad [2.40]$$

When we consider the case of no grating lobes, the absorption efficiency η_{absorp} becomes 100%, which is obtained by the substitution of equation [2.36] into equation [2.40]. This means that the infinite CMSA rectenna array can perfectly absorb the incident power under the condition of no grating lobes. This result is identical to that of a dipole antenna with a ground plane [ITO 84].

Figure 2.21 shows the absorption efficiency versus element spacing in the case of a square lattice ($L_x = L_y = L$). When the grating lobe is generated, the efficiency approaches zero because all the incident power flow is along the *x–z* plane or in the direction of propagation of the grating lobe. This phenomenon cannot be observed for an infinite array of dipoles with a ground plane in which the cancellation occurs between dipoles and their image [ITO 86]. However, the above discussion ignores the effect of the thickness of the substance. In fact, it is predicted that the efficiency is greater than zero because a surface wave generates before the formation of grating lobes.

Figure 2.22 shows the absorption efficiency in consideration of ohmic and dielectric losses of the CMSA, where the relative dielectric constant of the substrate is 2.6

and $\tan\delta$ is 2.2 × 10^{-3}. The greater spacing provides a smaller efficiency.

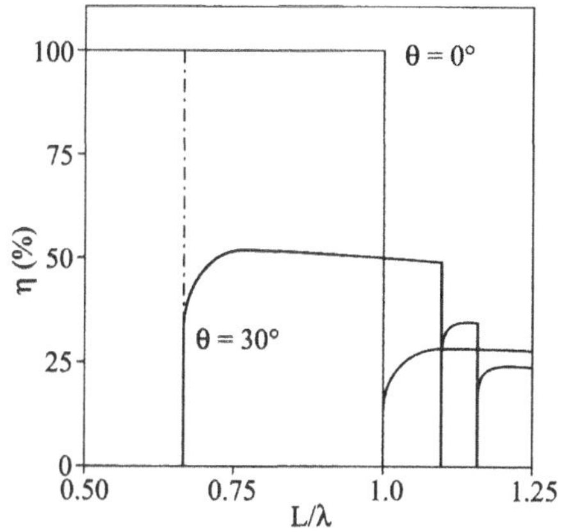

Figure 2.21. *Absorption efficiency versus element spacing, where the parameter is the incident angle θ ($L_x = L_y = L$) [ITO 86]*

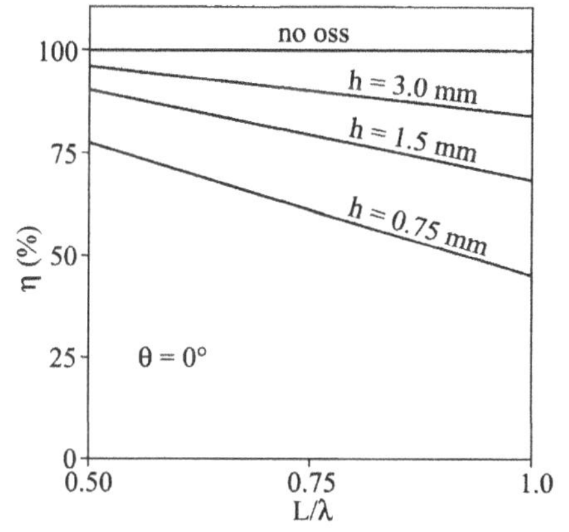

Figure 2.22. *Absorption efficiency versus element spacing, where CMSA possesses ohmic and dielectric losses [ITO 86]*

Otsuka *et al.* expanded the theory of the absorption efficiency of the infinite rectenna array to that of the finite rectenna array and conducted WPT experiments to confirm the absorption efficiency of the rectenna array [OTS 90]. An experimental setup and image of the rectenna array are shown in Figure 2.23. The experimental results indicate that the absorption efficiency of the rectenna array can be approximately 100% (Figure 2.24). On the basis of these studies, it can be considered that the BE can be almost 100% with $\tau > 2$ in the WPT system via radiowaves.

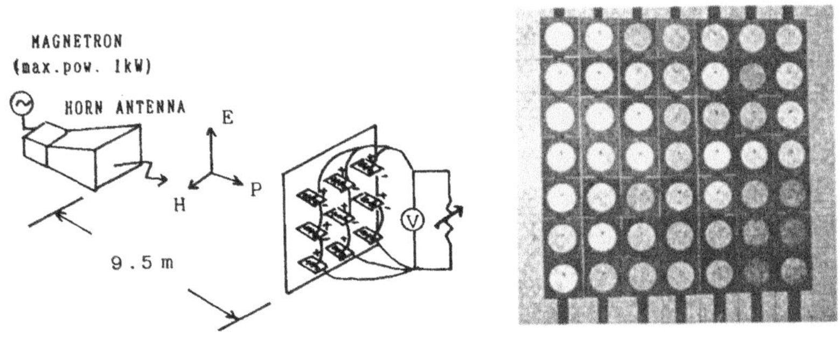

Figure 2.23. *Experimental setup of the investigation of the absorption efficiency of a rectenna array and an image of the rectenna array [OTS 90]*

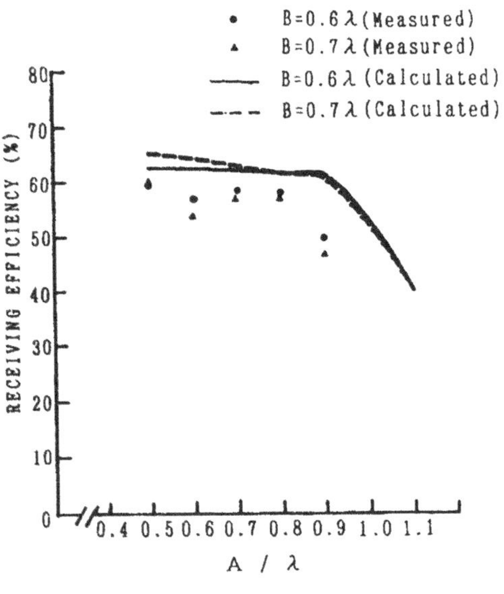

Figure 2.24. *Element spacing (x-direction) and receiving efficiency of a rectenna array [OTS 90]*

Chapter 3

Technologies of WPT

3.1. Introduction

As shown in Chapter 2, a power can be transmitted through radiowaves in accordance with the theory based on Maxwell's equations and antenna theory and also through high frequencies based on electromagnetic theory and coupling theory. In addition, a frequency converter with high efficiency is required to convert DC or 50/60 Hz AC to high frequencies for transmitting wireless power, and high-efficiency rectifiers are needed to convert from high frequency back to DC or 50/60 Hz AC.

Especially for microwave power transfer (MPT), high-power amplifiers (HPAs) using semiconductors and a microwave tube are often used as a power amplifier and microwave generator. In general, the characteristics of an HPA using semiconductors and the microwave tube are as follows:

– HPA using semiconductors (GaAs, GaN, etc.) for MPT (Figure 3.1(a)):

- low power (less than several dozen watts using GaAs semiconductors and less than several hundred watts using GaN semiconductors);

- low voltage requirement (<5 V using GaAs semiconductors and 50 V using GaN semiconductors);

- efficiency (<80% in the 2 GHz band, <70% in the 5 GHz band);

- small size (with monolithic microwave integrated circuit (MMIC) technology);

- expensive (depending on mass production);

- suitable for a phased array (because of low power and small size).

– Microwave tube (magnetron, klystron, traveling wave tube (TWT), etc.) for MPT (Figure 3.1(b)):

- higher power (greater than a kilowatt);

- higher voltage requirement (greater than several dozen kilovolts);

- efficiency (<80% in the 2 GHz band);

- larger size (necessary for vacuum);

- cheaper (cooker-type magnetron);

- unsuitable for a phased array (because of higher power and larger size with suppression of grating lobes).

The characteristics of power versus frequency for microwave tubes and HPA using semiconductors are shown in Figure 3.2.

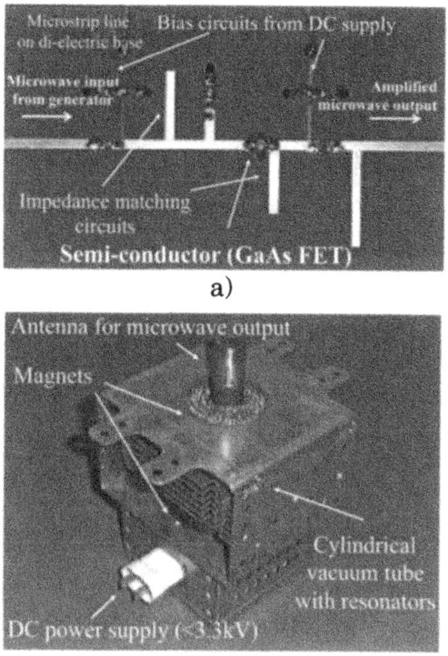

Figure 3.1. *a) HPA circuit using semiconductors (GaAs FET); b) cooker-type magnetron (microwave tube) for MPT*

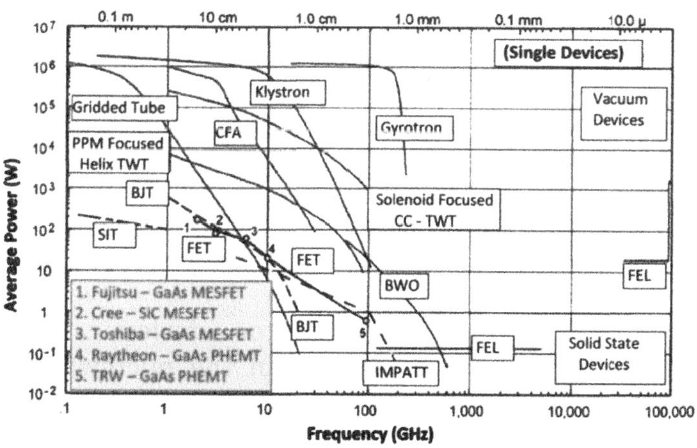

Figure 3.2. *Average RF output power versus frequency for various microwave tubes and semiconductors [TRE 02]*

In the 1960s, microwave tubes were often used for MPT experiments because the efficiency of the microwave tube was higher than that of an HPA using semiconductors. Recently, the HPA using semiconductors is sometimes used for MPT experiments using a phased array because the efficiency of the HPA is now much improved. The microwave tube is still used for point-to-point or short-distance MPT experiments without any beam forming. The cooker-type magnetron in particular is still often used because of its low cost. We should choose the HPA using semiconductors or the microwave tube depending upon the purpose and system design. Frequency converters using semiconductors in inductive coupling and resonant coupling wireless power transfer (WPT) systems are always used. In particular, an inverter circuit is used at lower frequencies.

The rectifying circuit is another important system for WPT. Electrical products in house or in industry typically require 50/60 Hz AC or sometimes DC. Therefore, the WPT system must convert high frequencies in the kilohertz to gigahertz range to DC or 50/60 Hz AC. However, it is quite difficult to convert high frequencies to 50/60 Hz AC, and hence a rectifier that directly converts kilohertz to gigahertz to DC is always used in a WPT system. In the rectifier, semiconductor diodes are often used. In an MPT system, a receiving system consisting of an antenna and the rectifier is called a rectenna, i.e. a rectifying antenna.

3.2. Radio frequency (RF) generation – HPA using semiconductors

After the 1980s, semiconductor devices, especially transistors, played a leading role in the microwave world rather than microwave tubes. This change was brought about by advances in mobile phone networks. Semiconductor devices are expected to expand microwave applications, e.g. the phased array and active integrated antenna, because of their manageability and capacity for mass production. After the 1990s, some MPT experiments were conducted in Japan with phased arrays using semiconductor HPAs.

Typical transistor devices for microwave circuits are the field effect transistor (FET), the heterojunction bipolar transistor (HBT) and the high electron mobility transistor (HEMT). Present materials for semiconductor devices are Si for frequencies below a few gigahertz and GaAs for higher frequencies. Recently, GaN and SiC are expected to be applied to microwave circuits. GaN and SiC are designated as wide-band-gap semiconductors, which can accommodate high power and high frequencies similar to GaAs semiconductors. At microwave frequencies, GaN is typically used. SiC is used for the inverter circuit at lower frequencies. The characteristics of output power versus frequency for modern GaN HPAs are shown in Figure 3.3.

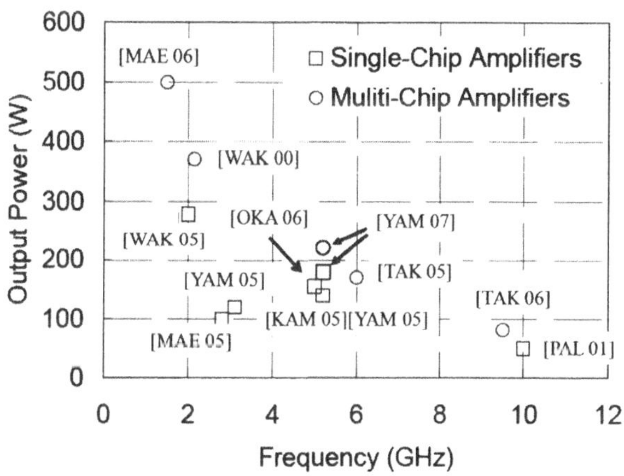

Figure 3.3. *Characteristics of output power versus frequency for modern GaN HPAs*

Microwave circuits are designed with these semiconductor devices. It is easy to control phase and amplitude via microwave circuits using semiconductor devices, e.g. amplifiers, phase shifters, modulators, etc. For microwave amplifiers, circuit design theoretically determines the efficiency and gain. Class A, B and C amplifiers are classified by the bias voltage used in the device. The amplifier circuit shown in Figure 3.4(a) is the same for class A, B and C amplifiers. These classes are also applied to kHz–MHz systems. The only difference is the operational point of V and I – bias voltage and the current, respectively – of a semiconductor. Theoretical amplifier efficiency is 50% for class A, 78.5% for class B and <100% for class C. In class B and C, efficiency is better than that in class A; however, the waveform is distorted and wave linearity cannot be maintained. For class C, the output power tends to be near zero when the efficiency is approximately 100%. As a result, class B and C devices are not typically used for wireless communication systems that require high wave linearity.

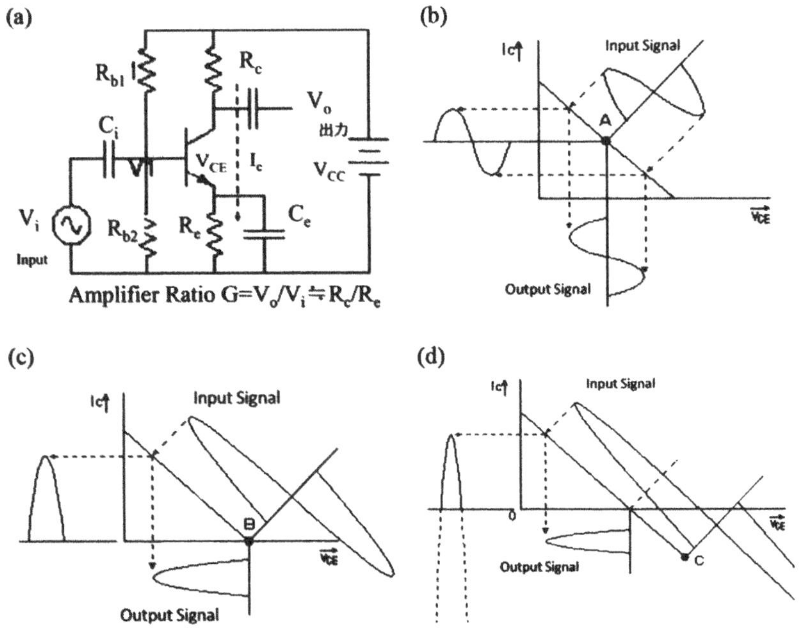

Figure 3.4. *Class A, B and C amplifiers. a) Circuit, b) operation point for class A, c) operation point for class B and d) operation point for class C*

Higher harmonics are effectively used to increase efficiency to levels that are theoretically 100% for class D, class E, class F and class inverse-F amplifiers operating at microwave frequencies. Class F [TYL 58, COL 99] and inverse-F [COL 99] amplifiers are especially expected to function as high-efficiency amplifiers for MPT systems. As shown in Figure 3.5, special impedance matching circuits controlling higher harmonics are added at the output and a transistor is driven at class B bias [HON 11]. The drain current becomes a half-wave-rectified waveform and higher harmonics occur at the class B bias circuit. The drain current $I_d(t)$ is transformed by the Fourier transform as follow1s:

$$I_d(t) = \frac{I_{\max}}{\pi} + \frac{I_{\max}}{2}\cos\omega t + \frac{2I_{\max}}{2\pi}\cos 2\omega t + \frac{2I_{\max}}{15\pi}\cos 4\omega t + \cdots + \frac{2I_{\max}}{(1-n^2)\pi}\cos\frac{n\pi}{2}\cos n\omega t \quad [3.1]$$

$$V_d(t) = V_{DC} - V_1\cos\omega t + V_3\cos 3\omega t + V_5\cos 5\omega t \cdots \quad [3.2]$$

where n = 2, 4, 6...

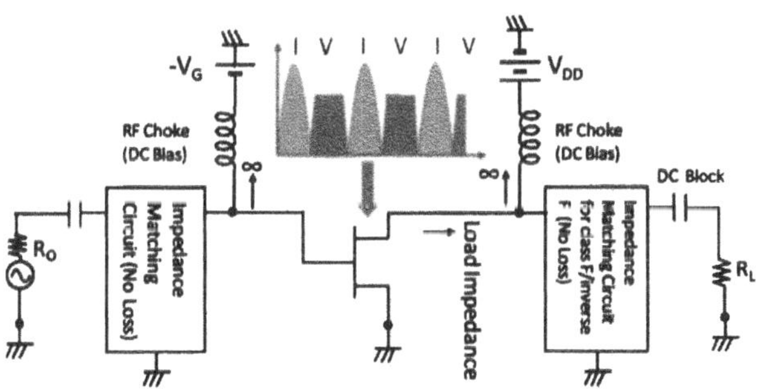

Load Impedance	Class F Condition	Class Inverse-F Condition
Fundamental Frequency f_0	Matched	Matched
Even Mode Higher Harmonics, $2f_0$, $4f_0$..	0 (Current Only)	∞ (Voltage Only)
Odd Mode Higher Harmonics, $3f_0$, $5f_0$..	∞ (Voltage Only)	0 (Current Only)

Figure 3.5. *A class F/inverse-F amplifier [HON 11]*

When the drain voltage $V_d(t)$ consists of DC, the fundamental frequency f_0, whose phase is 180° different from that of the fundamental frequency of the drain current, and odd harmonics, power consumption at the harmonics approaches zero and a power whose power factor is −1 occurs at the fundamental frequency. This means that DC power is converted to f_0 with theoretically 100% conversion efficiency if V_1 is optimized by an impedance matching circuit for f_0 and the bias voltage is adjusted. The relationships between equations [3.1] and [3.2] are shown in Figure 3.5.

The theoretical maximum efficiency and the number of harmonics of the current and voltage are presented in Table 3.1. It is important to increase the number of harmonics for higher efficiency. The theoretical efficiency of a class B amplifier is 78.5%. We should take care that an infinite number of even harmonics are required to realize a perfect half-wave-rectified waveform.

	f_0	$3f_0$	$5f_0$	Infinity
f_0	50% (Class A)	57.7%	60.3%	63.7%
$2f_0$	70.7% (Real class B)	81.7%	85.3%	90.3%
$4f_0$	75.0%	86.6%	90.5%	95.5%
Infinity	78.5% (Ideal class B)	90.7%	94.8%	100%

Table 3.1. *Relationship between harmonics and efficiency [HON 11]*

It is difficult to design the special impedance matching circuit with which we can control the higher harmonics. It is convenient to use distributed lines to design class F amplifiers, as shown in Figure 3.6. Some class F amplifiers have been developed for MPT applications. Figure 3.7

describes a class F amplifier operating at 1.9 GHz using GaN HEMT [ZHE 07]. The drain efficiency is 80.1% and power-added efficiency (PAE) = $(P_{out} - P_{in})/P_{DC})$ is a maximum of 72.6% at 1.9 GHz. The same group developed a class F amplifier operating at 5.65 GHz using AlGaN/GaN HEMT [KAM 11]. The drain efficiency is 90.7% and PAE is a maximum of 79.5% and the saturated power is 33.3 dBm.

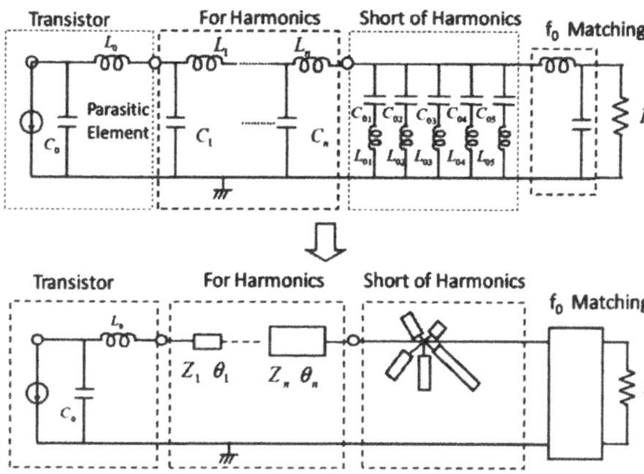

Figure 3.6. *Class F amplifier using distributed lines [HON 11]*

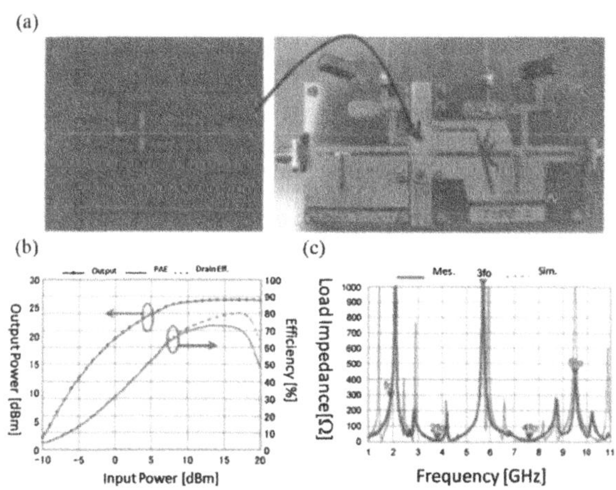

Figure 3.7. *Class F amplifier operating at 1.9 GHz using GaN HEMT developed by Univ. of Electro-Communications in Japan [ZHE 07]*

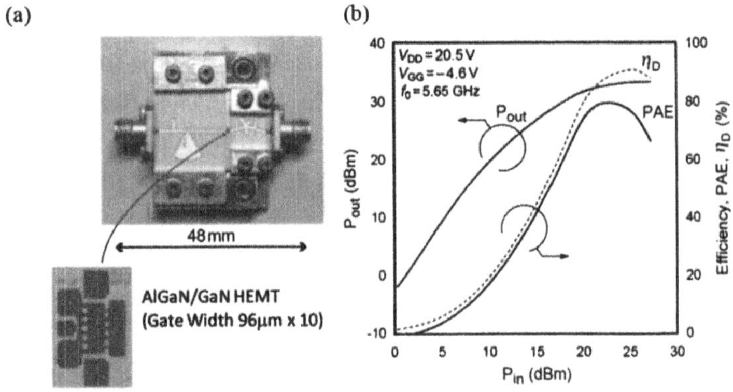

a) b)

Figure 3.8. *Class F amplifier operating at 5.65 GHz using AlGaN / GaN HEMT developed by Univ. of Electro-Communications in Japan [KAM 11]*

The other requirement of semiconductor amplifiers for MPT use is amplifier linearity because the power level for MPT is much higher than for wireless communication systems. In addition, we have to suppress unexpected spurious radiation to reduce interference. The maximum efficiency is usually realized at a saturated bias voltage. It does not guarantee linearity between input and output microwaves, and nonlinearity causes high spurious radiation, which must be suppressed in MPT. Therefore, dissolution of tortuous relationships between efficiency and linearity is expected for MPT. This is one of the reasons why class F/inverse-F amplifiers are suitable for MPT applications because harmonics are used to increase the efficiency.

3.3. RF generation – microwave tubes

Several microwave generators or amplifiers are in use, such as magnetron, klystron and TWT/traveling wave tube amplifier (TWTA). Each microwave tube is typically used for cooking, plasma heating and satellite communications, respectively. Since the 1960s microwave tubes have been used for MPT.

3.3.1. *Magnetrons*

A magnetron is a crossed field tube in which $\mathbf{E} \times \mathbf{B}$ forces electrons emitted from a cathode to take a cyclonical path to an anode. It is a self-oscillatory device in which the anode contains a resonant RF structure. Figure 3.9 describes the structure of a magnetron and the principle of microwave generation in the magnetron. Efficiency of microwave generation in the magnetron is theoretically very high. High voltage is used and electrons, which flow in a vacuum, serve as the microwave source. The frequency is tuned only by the shape of a resonant cavity. The efficiencies of commercial magnetrons are shown in Figure 3.10.

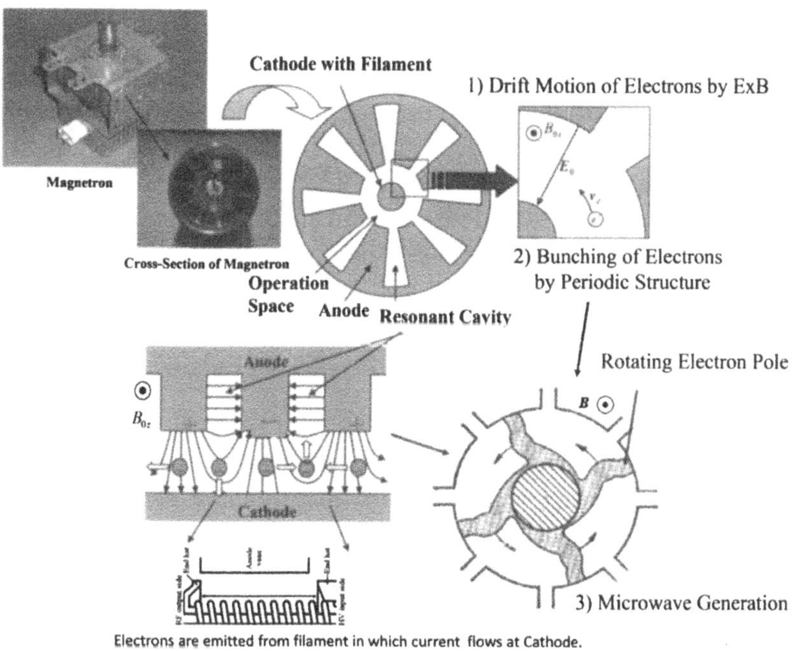

Figure 3.9. *Structure of the magnetron and the principle of microwave generation in a magnetron*

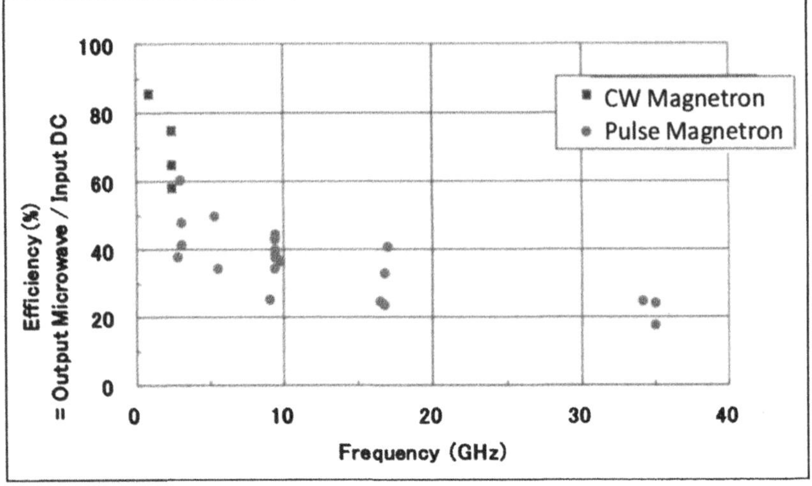

Figure 3.10. *Efficiency of a magnetron*

The magnetron has a long history since its invention by A.W. Hull in 1921. The practical and efficient magnetron tube gathered world interest only after K. Okabe proposed the divided anode-type magnetron in 1928. Magnetron technologies were advanced during World War II, especially in the Japanese army. In 1945, P.L. Spencer of the Raytheon Company in the United States found that microwaves were emitted from the magnetron of a radar-melted chocolate in his pocket. Raytheon thus got a patent on the microwave oven and began producing ovens in 1952. After this, magnetrons were mainly produced for microwave ovens.

As a result, magnetrons in the 500–1,000 W range are widely used in microwave ovens operating at a frequency of 2.45 GHz, and it is a relatively inexpensive oscillator (less than $5). Magnetrons can be produced mainly using metal frames and magnets, and can therefore be produced at low cost. There is a net global capacity of 45.5 GW/year for all magnetrons used in microwave ovens, which have a yearly production of 50–55 million. The history of the magnetron is the same as that of the microwave oven. The first microwave oven using a magnetron sold in the United States shortly after World War II for more than $2,000, which is equivalent to approximately $20,000 today. In the 1960s, Japan played an important role in reducing the cost of the microwave oven. Compared with American manufactured tubes, which cost approximately $300 in the 1960s, the Japanese manufactured tubes cost less than $25. In 1970, U.S. manufacturers sold 40,000 ovens at $300–$400 a piece; however, by 1971, the Japanese had begun exporting low-cost models priced at $100–$200 less. Sales rapidly increased over the next 15 years, rising to approximately 1 million by 1975 and 10 million by 1985, nearly all of them manufactured by the Japanese. However, history repeats itself. Instead of Japanese microwave ovens, Korean- and Chinese-manufactured ovens presently are responsible for the reduction in the cost of the microwave ovens. Figure 3.11 shows a comparison of the market prices of semiconductor amplifiers and magnetrons in 2012. Magnetrons are overwhelmingly cheaper than semiconductor amplifiers.

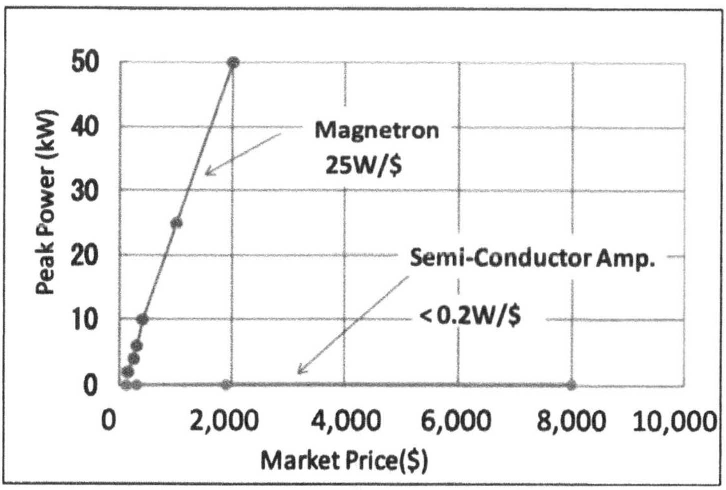

Figure 3.11. *Comparison of market price of semiconductors and magnetrons*

However, owing to our recent study, we know that the output microwave spectrum of the cooker-type magnetron depends on the stability of the DC power source and filament current [KAM 05]. If a stabilized DC power source is used for the magnetron and the filament current is deactivated after stable oscillation occurs, the spectrum, including low frequency and high frequency, is quiet, pure and adequate for its use in MPT (Figure 3.12). The Q factor of the magnetron with a stabilized DC power source is approximately 1.1×10^5. Peak levels of higher harmonics are below –60 dBc, and other spurious radiation is below –100 dBc.

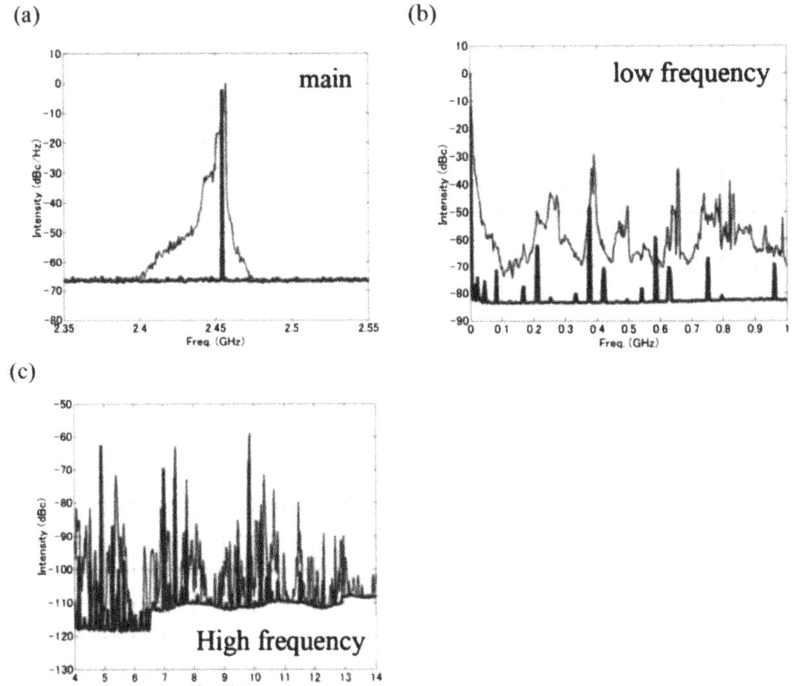

Figure 3.12. *Output spectrum of a commonly used magnetron (bold: with a stabilized DC power supply and filament off, normal: with half-wave voltage doubler power supply and filament on). a) Main frequency; b) low frequency; c) high frequency*

There remains, however, the problem of the frequency shift caused by magnetron heating and the impossibility of phase/frequency control even if the spurious radiation is suppressed. In the 1960s, Dr. Brown advanced the technologies used in magnetrons and developed a magnetron amplifier [BRO 81, BRO 89] whose frequency/phase was stabilized and controllable. For the magnetron amplifier, he adopted injection locking and a feedback loop to a magnetic field. Furthermore, he proposed the application of a magnetron amplifier to an MPT and solar power satellite (SPS).

Injection locking is accomplished by operating the magnetron as a reflection amplifier [SIV 94]. It can be used in a phased array using semiconductor microwave generators [YOR 98, STE 87]. The effect was formulated by R. Adler as Adler's equation in equation [3.3] [ADL 46].

$$\frac{\Delta f}{f} = \frac{2}{Q_E}\sqrt{\frac{P_i}{P_o}} \qquad [3.3]$$

where Δf is the locking frequency width, which is basically the difference between the frequency of an injected weak signal and the output frequency of the oscillator, f is the frequency of the injected signal, Q_E is an external Q factor for the oscillator, in this case the magnetron, and P_i and P_o are the input power for locking and the output power of the oscillator, respectively. Adler's equation indicates that the frequency of the oscillator is locked to the frequency of the injected weak signal if Δf, which is decided by Adler's equation, remains (see Figure 3.13). For example, an estimated Δf is shown in Figure 3.14. Experimental results for the injection locking of a magnetron are shown in Figure 3.15. As time passes, the frequency of the magnetron moves to the frequency of the injected signal. The bandwidth of the main frequency of the magnetron becomes narrower than that without injection locking.

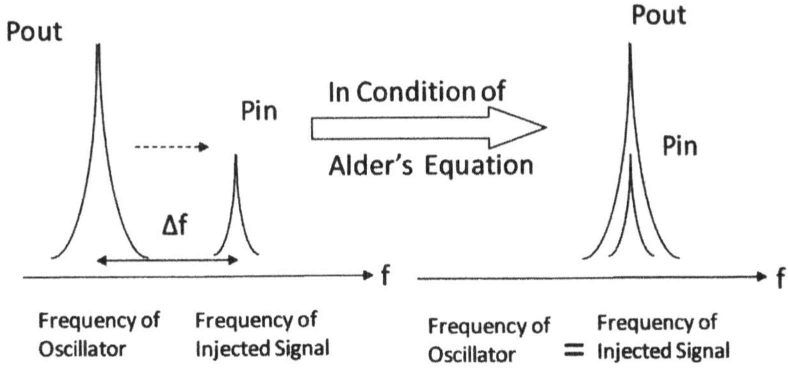

Figure 3.13. *A description of injection locking*

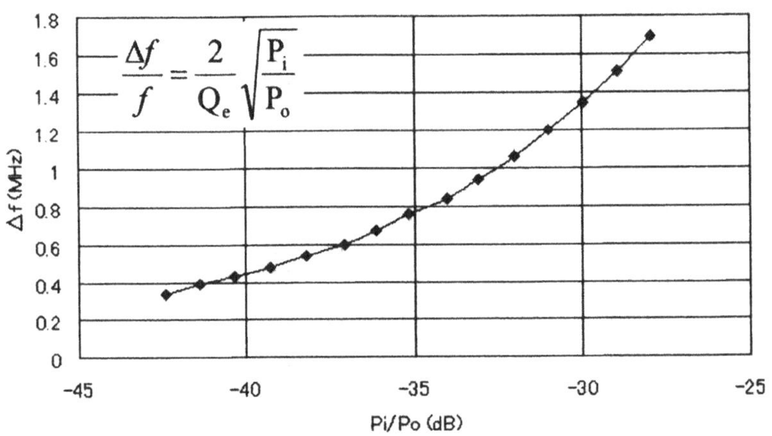

Figure 3.14. *Experimental result of the locking range* Δf

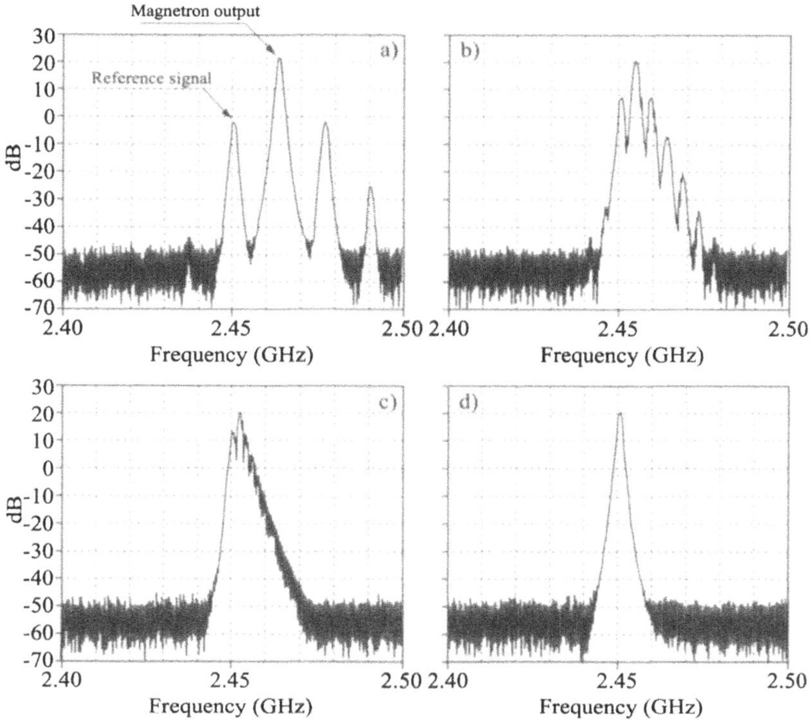

Figure 3.15. *Experimental results of injection locking of a magnetron*

Although the frequency can be locked and stabilized with the injection locking technique, the difference in phases remains even when the frequency is locked by injection locking. The difference in the phases θ between the injected signal and the oscillated signal is described by the following equation.

$$\theta = \sin^{-1} \frac{\Delta f' Q_E}{f\sqrt{P_i / P_o}} \quad \left(|\theta| \leq 90^o\right) \qquad [3.4]$$

where $\Delta f'$ is the original difference in the frequency between the injected signal and the oscillated signal before injection locking. The experimental results showing the phase difference θ are given in Figure 3.16.

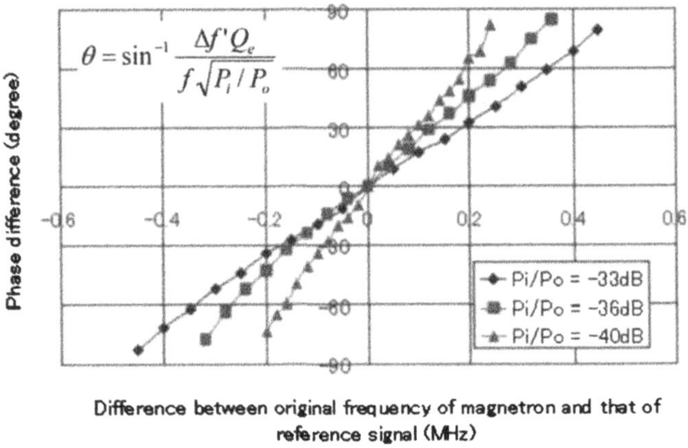

Figure 3.16. *Experimental results of the phase difference* θ

It is enough to use only one magnetron for MPT systems because the frequency of the magnetron is stabilized by injection locking. However, it is not enough to use two or more magnetrons for a phased array because the phases of the magnetrons are different. The frequency/phase of the magnetron can be controlled by the anode current (electric field **E**) or magnetic field **B**. Therefore, a phase-locked loop (PLL) feedback is used with injection locking in order to synchronize the phase of the reference signal and the microwave output. This is called the phase-controlled magnetron (PCM) developed by Kyoto University or a magnetron amplifier developed by Brown.

Kyoto University's group has developed a PCM for MPT using different technologies from the one developed by Brown in the 1960s. For Kyoto University's PCM, a PLL feedback to the anode current is adopted (Figure 3.13) [SHI 01]. A block diagram of the PCM is shown in Figure 3.17 and an image of the device is shown in Figure 3.18 [SHI 03]. However, a PLL feedback to a magnetic coil was adopted for the magnetron amplifier (Figure 3.19 and Figure 3.20) [BRO 89]. A phase and an amplitude control of the magnetron were proposed whose block diagram is shown in Figure 3.21 [BRO 81]. A phased array has been developed using PCMs by Kyoto University [SHI 03, SHI 00] (see section 3.4.4). Some research groups attempted to develop new magnetron systems in which they could control the phase of the microwave emitted from a cooker-type magnetron [HAT 98, HAT 99, CEL 97, SAN 00, TAH 05, TAH 06]. In these developed magnetrons, injection locking and a PLL feedback were adopted in the same manner as adopted in Brown's work.

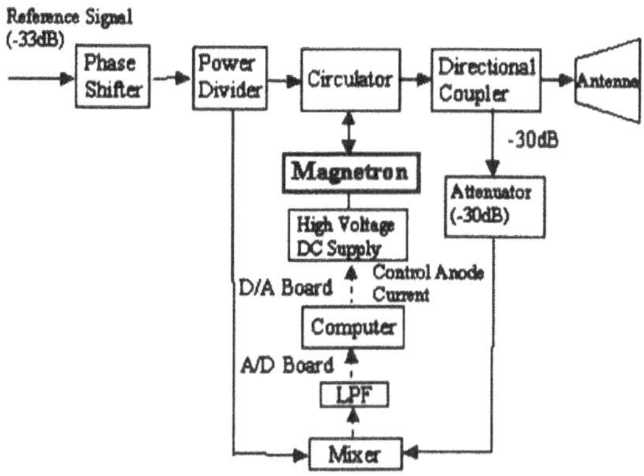

Figure 3.17. *Phase-controlled magnetron system with an anode current feedback by Kyoto University [SHI 01]*

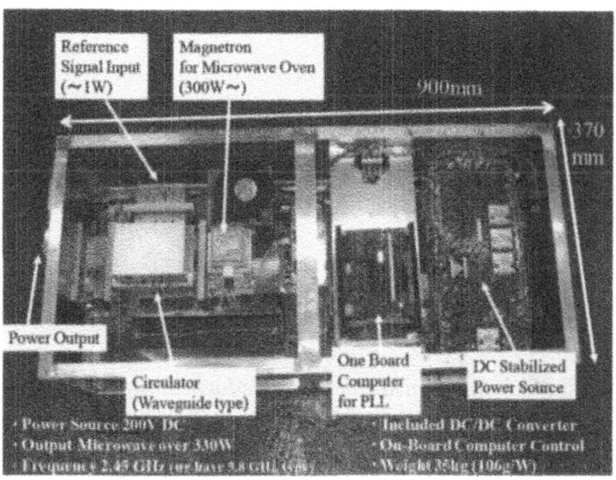

Figure 3.18. *Image of the phase-controlled magnetron [SHI 03]*

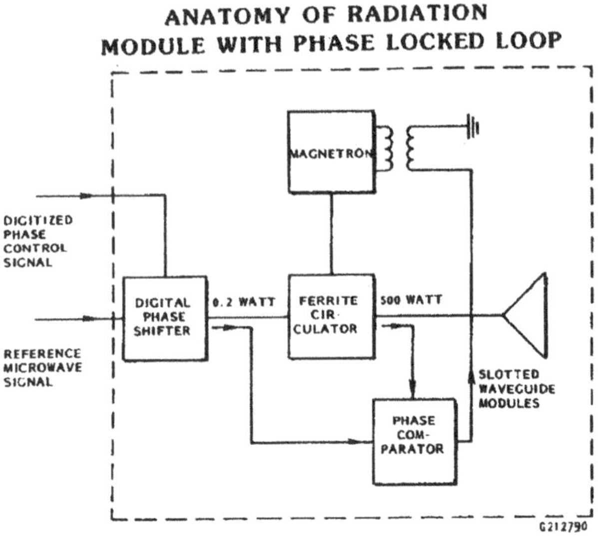

Figure 3.19. *Circuit for phase-locked, high-gain magnetron directional amplifier by Brown [BRO 89]*

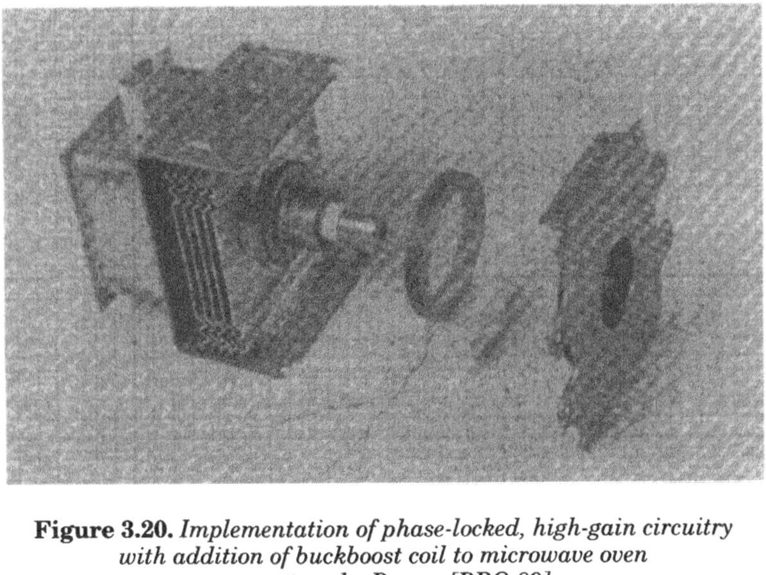

Figure 3.20. *Implementation of phase-locked, high-gain circuitry with addition of buckboost coil to microwave oven magnetron by Brown [BRO 89]*

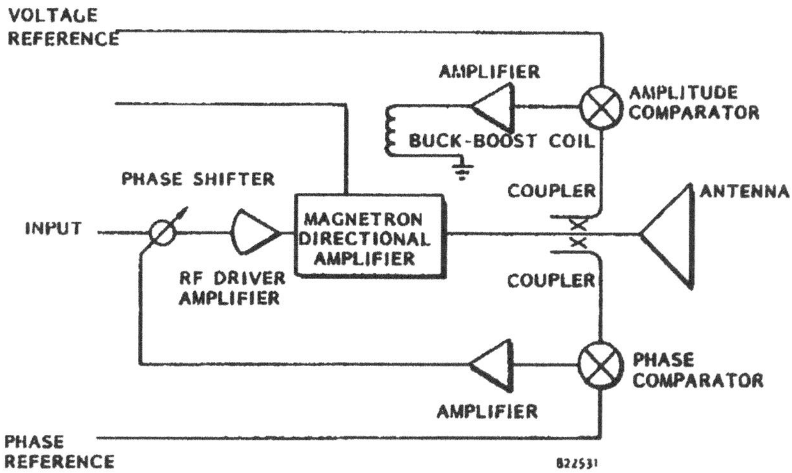

Figure 3.21. *Phase- and amplitude-controlled magnetron system with magnetron amplifier by Brown [BRO 81]*

PCMs can be used and have been used for phased arrays; however, some problems remain. One problem is a power loss of more than 10% in the circulator to inject the reference signal to the magnetron for injection locking. The other problem is the impossibility of power control while controlling the phase/frequency of the magnetron, which is theoretically due to the magnetron itself. To solve these two problems, a PLL magnetron and phase- and amplitude-controlled magnetron (PACM) have been developed by Kyoto University.

It is possible to stabilize the microwave of the magnetron by only using PLL feedback without injection locking. This is done by using a PLL magnetron. In the PLL magnetron, the magnetron and a DC power source are considered as voltage-controlled oscillators in normal PLL. A block diagram of the PLL magnetron is shown in Figure 3.22. Figure 3.23 is equivalent to the block diagram shown in Figure 3.22. An open-loop transfer function of the PLL magnetron can be estimated. The low pass filter (LPF) transfer function $F_{LPF}(s)$ is defined as follows:

$$F_{LPF}(s) = \frac{1 + sC_1R}{s(C_1 + C_2)(1 + s\frac{C_1C_2R}{C_1 + C_2})} \quad [3.5]$$

The high-voltage source is considered as a first delay, which is defined as follows:

$$F_{pw}(s) = \frac{1}{1 + sT_{pw}} \quad (T_{pw} = 1.59 \text{ s}) \quad [3.6]$$

As a result, the open-loop transfer function F_{OPloop} can be written as follows:

$$F_{OPloop}(s) = \frac{K_{VCO}K_{PC}F_{LPF}(s)}{s(1 + sT_{pw})N_T} \quad [3.7]$$

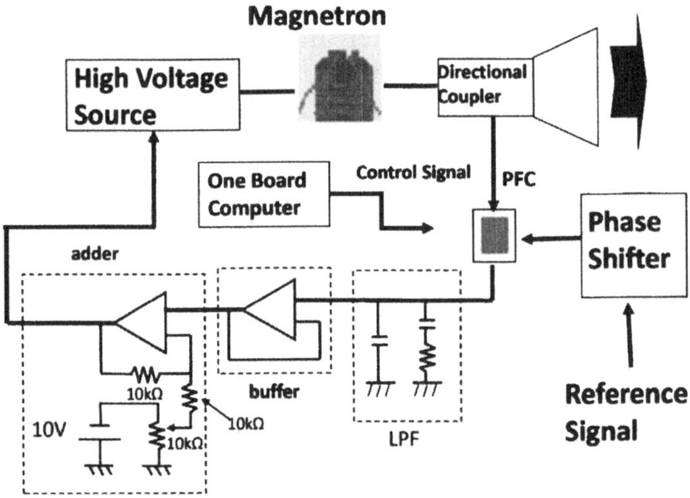

Figure 3.22. *Diagram of the PLL magnetron*

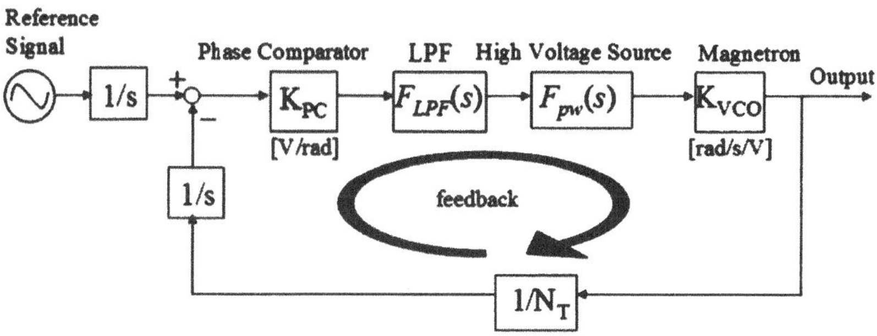

Figure 3.23. *Block diagram for estimation of the open-loop transfer function*

With equation [3.7], stability of the loop can be estimated with bode diagrams, as shown in Figure 3.24. The phase margin is approximately 40° and the feedback system is stable. Figure 3.25 shows the experimental results of phase and frequency locking of the PLL magnetron. The phase and frequency of the magnetron are stabilized and controlled by the PLL feedback loop only. The frequency stability is approximately 10^{-5} to 10^{-6}. It is sufficient to control the microwave beam power, which has a lower stability than that of the PCM with injection locking.

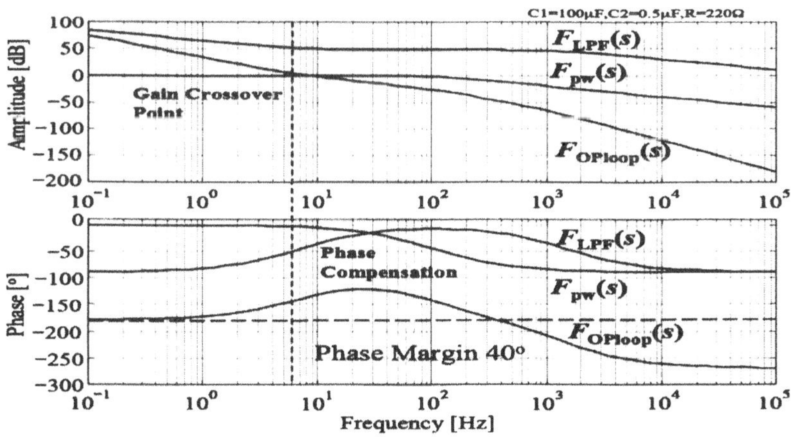

Figure 3.24. *Bode diagrams of PLL feedback shown in Figure 3.22*

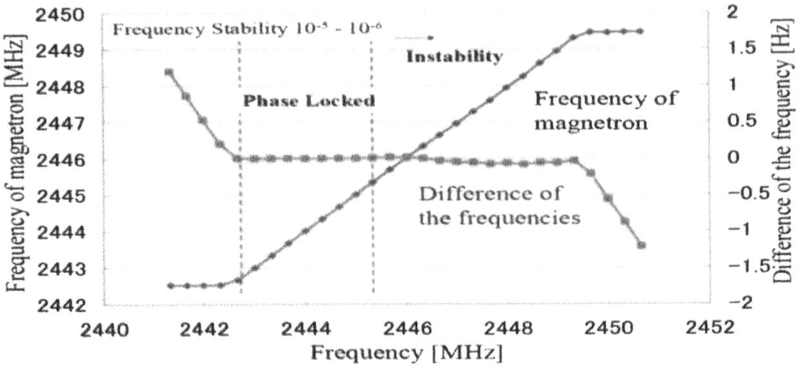

Figure 3.25. *Experimental results of phase and frequency locking of the PLL magnetron*

As mentioned above, the frequency/phase of the magnetron can be controlled by the anode current (electric field **E**) or magnetic field **B**. The problem is that the microwave power is simultaneously controlled by the anode current or the magnetic field. When we control and stabilize the frequency/phase of the magnetron by controlling the anode current, the microwave power is simultaneously controlled to unexpected levels. To solve this problem, a PACM with two feedback loops to the anode current and the magnetic field has been developed at Kyoto University. A block diagram of the PACM is shown in Figure 3.26. Figure 3.27 shows the results of amplitude control of the PACM. The amplitude of the microwave from the PACM is controlled from 300 to 500 W, with a stability of within 1% of the phase. Figure 3.28 shows the time dependence of the phase and the amplitude of the microwave from a magnetron, the PCM and the PACM. The frequency stability and error in phase and amplitude of the PACM are less than 10^{-6}, within 1° and within 1%, respectively.

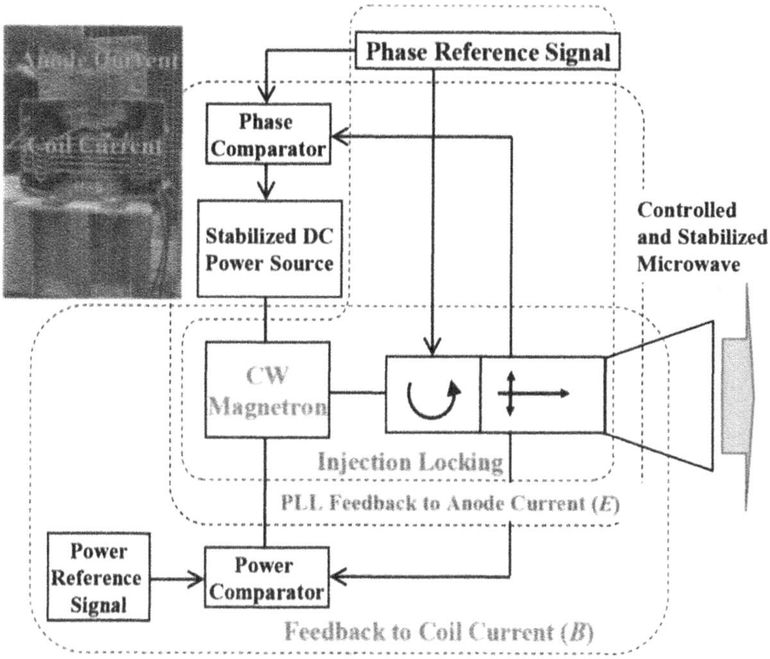

Figure 3.26. *Block diagram of the PACM*

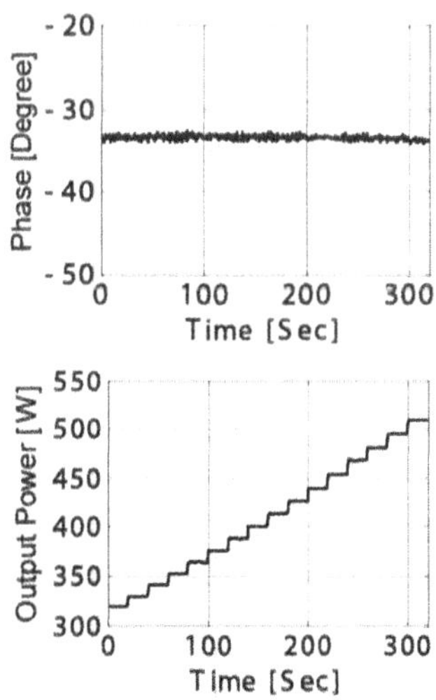

Figure 3.27. *Power control of the PACM*

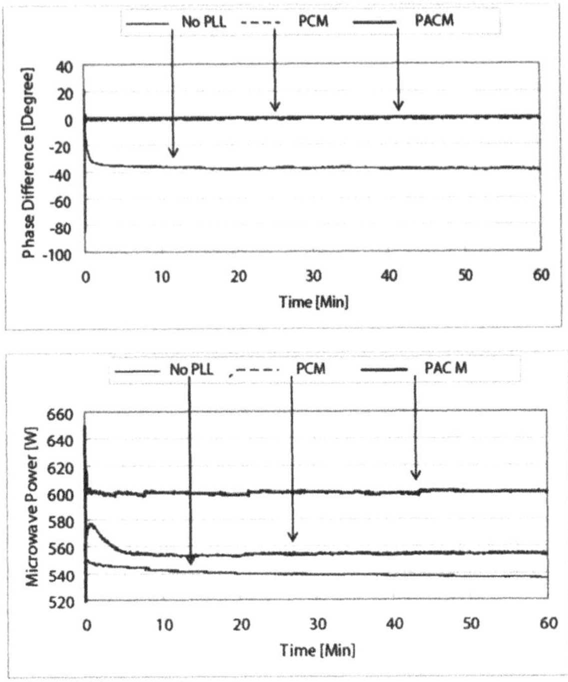

Figure 3.28. *Time dependence of the phase and amplitude of the microwave from a magnetron, the PCM and the PACM*

3.3.2. *Traveling wave tube/traveling wave tube amplifier*

TWT was invented by Kompfner during World War II and advanced theoretically and improved by Pierce and Field in 1945. The TWT is a linear beam tube with a helix structure. The helix slow wave structure (SWS) slows the RF waves down to just below the velocity of the electron beam. In the TWT, the interaction between the RF waves and the electron beam is continuous along the length of the SWS. The TWT can be used for an amplifier and is called a TWTA in that capacity. The longer the beam tube, the higher the gain. The applied frequency of the TWTA is very wide, ranging from 1 GHz band to the 60 GHz band. The typical output power of the TWT is a few hundred watts.

The TWTA is widely used in television broadcasting satellites and communication satellites. The TWTA has a proven track record in space. Before the 1980s, the efficiency of the TWTA was very low, approximately 30%. As such, this is not sufficient for use in an MPT system. There have been no MPT system designs or experiments using conventional TWTA. However, in recent years, a TWTA using a technique called velocity tapering energy recovery has been developed [GRA 99]. In this way, the net conversion rate has risen to approximately 70% [HEI 97] (Figure 3.29). The market for the TWTA has grown since 1972, and its cost has decreased [HEI 97]. Heider [HEI 97] identifies the main reasons for this price decrease as: (1) reduction in development time and effort due to standardization of the product, (2) reduction in parts cost due to higher volume purchase, (3) reduced manufacturing costs by manufacturing increased numbers of TWTAs in a shorter time frame and by more automatization in the manufacturing process and (4) reduction in test time and efforts due to the higher credibility of the product.

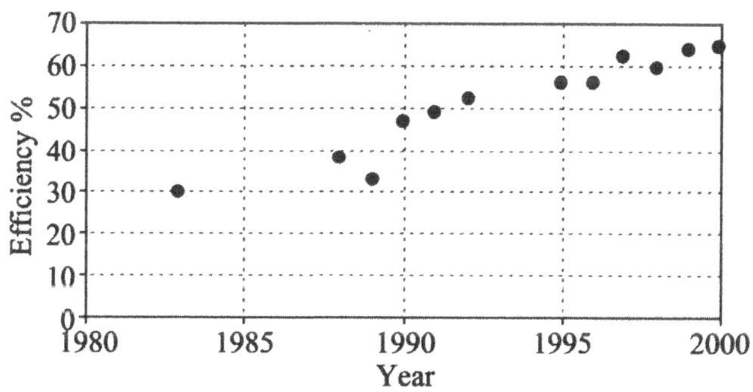

Figure 3.29. *Trend of DC–RF conversion efficiency for TWTA [HEI 97]*

Trends in the development of the TWT are microwave power module (MPM) and the phased array TWT. The MPM combines the best aspects of TWT, semiconductor amplifiers and state-of-the-art power supply technology into one package. This makes MPM a good candidate for space applications because it has high conversion efficiency, small size and low weight. In the near future, MPT systems using TWTA could be considered.

The TWTA is normally used as one transmitter. However, a phased array using TWTAs are tested for broadcasting satellite systems [YAM 07b]. TWTA array is used as a feed array of a parabolic antenna. For the phased array, low power TWTA is required. Mini-TWTA is developed for the phased array and its power, gain and efficiency are 10.8 W, 36.3 dB and 46.2% at 21.4–22 GHz, respectively.

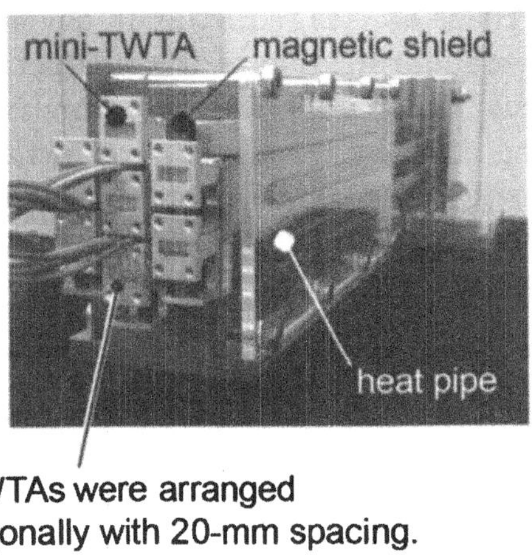

Figure 3.30. *Phased array using TWTAs [YAM 07b]*

3.3.3. *Klystron*

The klystron was invented by the Varian brothers in the late 1930s. A klystron is also a linear beam tube containing cavities. Electrons are emitted from the cathode and an electron beam passes through the cavities. When the RF is input from the input cavity, the electron beam is modulated and the RF input is amplified. A klystron is an HPA from tens of kilowatts to a few megawatts with high efficiency of more than 70%. However, it requires a strong power supply and a heavy magnet. Klystrons are used for broadcast applications in the 400–850 MHz band. They are also used for uplinks (earth stations beaming to orbital satellites). The other application of the klystron is in fusion.

The klystron was used in an MPT demonstration in 1975 at the JPL Goldstone Facility. One klystron transmitted a microwave of 450 kW at an operating frequency of 2.388 GHz. The klystron is suitable for large MPT systems such as SPS. The SPS designed by NASA/DOE in 1980 was designed with a phased array consisting of klystrons. However, no klystron phased array system has yet been constructed.

3.4 Beam-forming and target-detecting technologies with phased array

3.4.1. *Introduction*

We know that beam direction can be controlled in the phased array. However, the compatibility of high-efficiency beam forming of MPT is very difficult. In Japan, some trials of MPT experiments with a phased array were carried out in the 1990s and in the 21st Century. Phased array with semiconductors for the MPT in Japan are described, which were developed for the microwave lifted airplane experiment (MILAX) and the International Space Year – Microwave Energy Transmission in Space (ISY-METS) experiment in 1990s, and JAXA's SPS demonstrator, USEF/METI's active integrated antenna (AIA) system and corporative works with Mitsubishi Electric Cooperation and Kyoto University. Magnetron phased arrays, which were developed in Kyoto University, are also developed in Japan. Retrodirective, a target detection technique using radiowaves, is important for beam forming. The Japanese trials using a phased array and a retrodirective target-detecting system are described in this section. This section is based on [SHI 13C].

Each project of the phased array is independent. In each project, they had a research target, for example, to fly a fuel-free airplane with microwaves, to develop the thickest phased array or to develop a magnetron phased array itself. But they are based on the previous project to develop new phased array. The trend of the development of the phased array is shown in Figure 3.31. Key words are "higher frequency", "higher efficiency" and "thick size and lightweight". In each project, they adopted the newest technologies to develop the phased array.

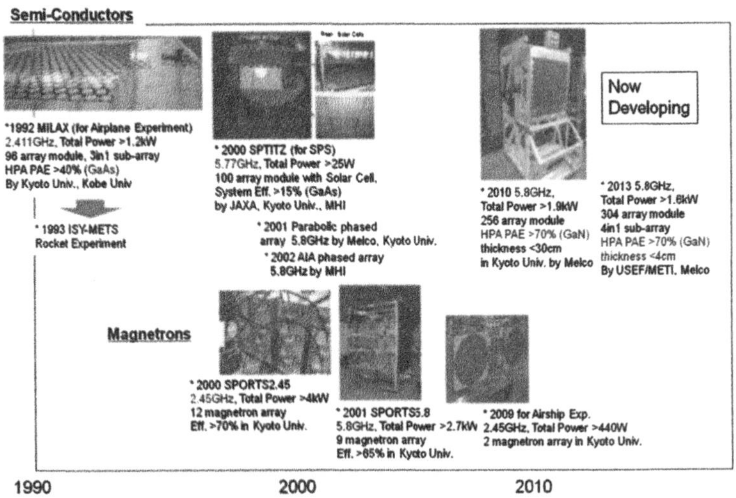

Figure 3.31. *Historical summary of phased array for MPT [SHI 13]*

3.4.2. *Phased array in the 1990s*

In 1983, Kyoto University's group in Japan carried out the first successful MPT rocket experiment called the microwave ionosphere nonlinear interaction experiment (MINIX). The experiment yielded new plasma data and theory about the nonlinear interaction between high-power microwave beams and ionospheric plasmas [MAT 86, KAY 86, NAG 86]. A cooker-type magnetron and waveguide antenna was used to transmit the 2.45 GHz microwave power from the mother transmitter rocket to the daughter receiver rocket in MINIX. Ten years later, they planned the next rocket MPT experiment with a phased array to describe detailed plasma physics. Data about the angle between the magnetic field and excited plasma wave were required to calculate the level of the excited plasma wave, which is caused by the high-power microwave beam. They also had to concentrate the microwave power to a narrow area to estimate the angle between a magnetic field and excited plasma wave. Therefore, the phased array was suitable to control the microwave beam direction and to calculate the angle between a magnetic field and excited plasma wave. The magnetic field could be measured in the rocket experiment. The microwave beam direction was estimated by the beam control data, together with the experimental data about beam direction collected before the rocket experiment. The second rocket MPT experiment with a phased array was called the ISY-METS by Kyoto University, Kobe University, Texas A&M University from USA, Communication Research Laboratory (CRL) (presently known as the National Institute of Information and Communications Technology (NICT)) and the Institute of Space and Astronautical Science (ISAS).

Before the ISY-METS experiment, the fuel-free airplane experiment called the MILAX was carried out in August 1992 [MAT 93]. The phased array developed for ISY-METS was used to transmit the microwave power. In total, 96 GaAs semiconductor amplifiers and 4-bit digital phase shifters were connected to 288 antenna elements at 2.411 GHz. This meant that there were three antennas in one amplifier subarray (Figure 3.32). The diameter of the phased array was approximately 1.3 m. The measured beam pattern is shown in Figure 3.33. The beam width was approximately 6°. The gain of the amplifier was 42 dB at 0 dBm input. Output power was approximately 42 dBm (Figure 3.34). The PAE of the amplifier was approximately 40%. The total microwave power was 1.25 kW and consisted of a continuous wave (CW) with no modulation. The measured power density is shown in Figure 3.35.

Figure 3.32. *First phased array for MPT in MILAX in 1992*

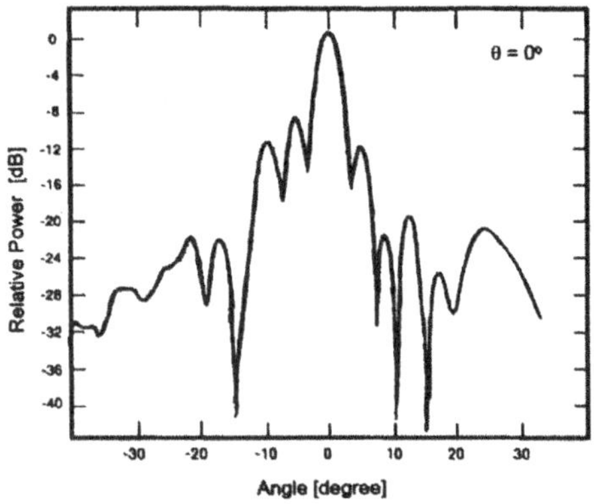

Figure 3.33. *Beam pattern of the MILAX phased array*

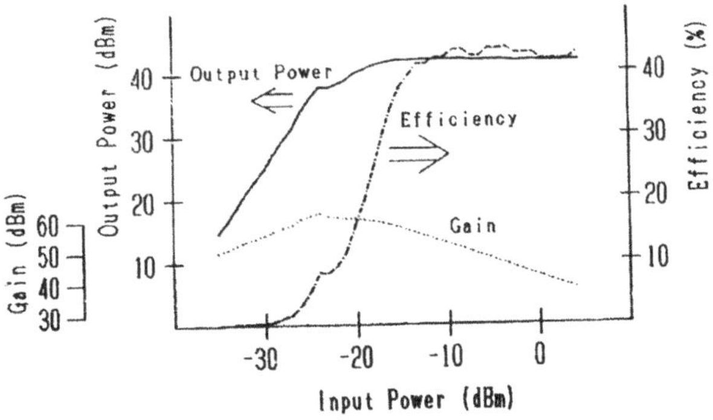

Figure 3.34. *Measured output power, efficiency and gain of the GaAs amplifiers in MILAX and ISY-METS*

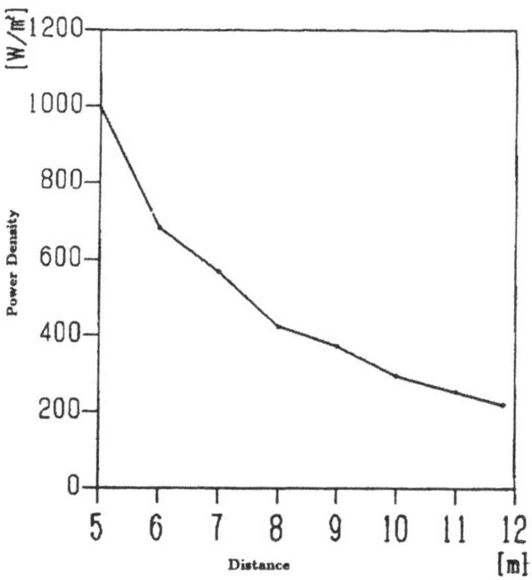

Figure 3.35. *Transmitted power density at different distance*

The phased array was assembled on the roof of a car. The car drove under the fuel-free airplane as much as possible and the microwave was controlled toward the fuel-free model airplane using a computer and data from two charged coupled device (CCD) cameras, which detected the position of the target. One of the CCD cameras is shown on the right-hand side of Figure 3.32. The rectenna array on the airplane's body is shown in Figure 3.36. There were in total 120 rectennas. The element spacing was 0.7λ. The efficiency of the rectenna was approximately 61% at 1 W output DC power. The airplane flew freely using only the microwave power that was fed by the phased array. The airplane flew at approximately 10 m above ground level. The maximum obtained DC power from the rectenna array was approximately 88 W and was sufficient to fly the airplane. This was the first MPT field experiment using a phased array in the world.

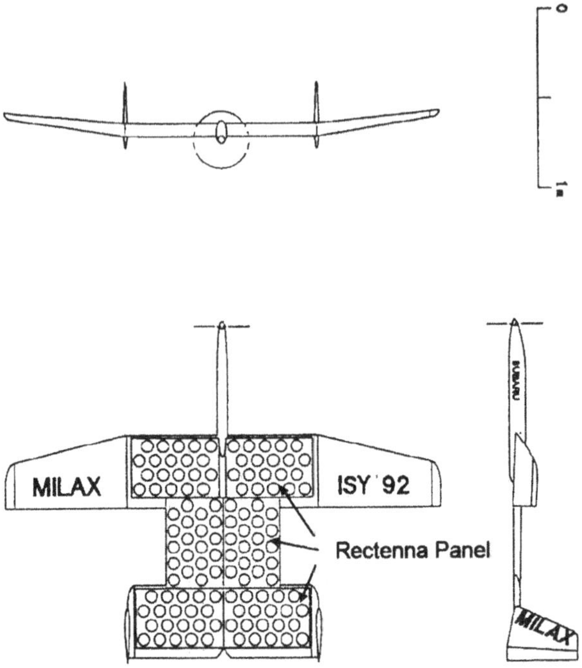

Figure 3.36. *MILAX airplane and rectenna array*

After the success of MILAX, the rocket experiment with a phased array called ISY-METS was carried out in February 1993 [KAY 93]. The phased array was arranged to cross plane and folded into the head cone of the rocket (Figure 3.37). There were ($2 \times 8 = 16$) antennas fitted to each of the four panels. They used the same GaAs semiconductor amplifiers and 4-bit digital phase shifters. Figure 3.38 shows the measured beam pattern that is concentrated on one point, 10 m away from the antenna, and there was no other documented condition to measure the beam pattern [KAY 93]. Thus, we should consider the vertical axis as relative power only. The large side lobe was achieved by the positioning of the antennas. Figure 3.39 shows the concentrated power received by a dipole antenna at different distances from the transmitting antenna by a solid line, while circles indicate the flight data [AKI 93]. They succeeded in the beam control in space from the phased array in the mother rocket to a receiver in the daughter rocket.

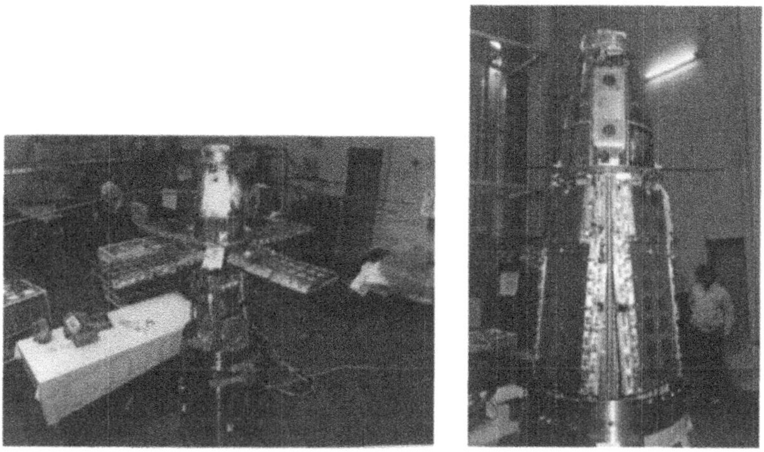

a) b)

Figure 3.37. *Phased array in ISY-METS rocket experiment. a) Expanded shape for experiment; b) folded shape in launch*

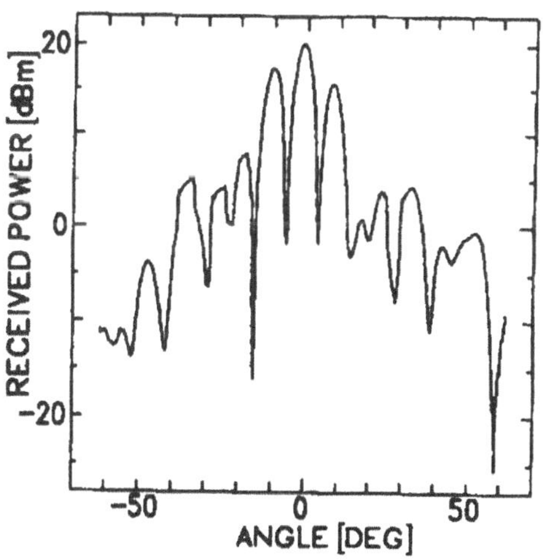

Figure 3.38. *Beam pattern of the ISY-METS phased array*

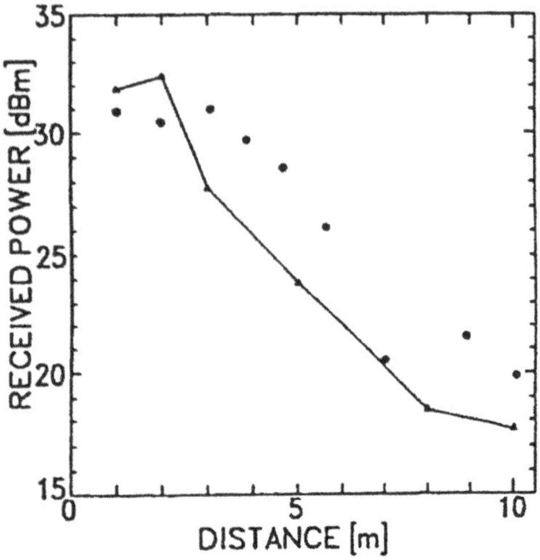

Figure 3.39. *Received power by a dipole antenna at different distances*

3.4.3. *Phased array in the 2000s*

The phased arrays in MILAX and ISY-METS were insufficient for power transfer. Recently, the phased array technology itself advanced into synthetic aperture radar (SAR). The related technology of adaptive arrays advanced into multiple input multiple output (MIMO) systems. However, MPT requires higher efficiency in the phased array.

In the 2000s in Japan, some trials were carried out as part of the development of high-efficiency phased arrays, mainly for SPS applications. One of the projects was SPRITZ in FY2000 (Figure 3.40) [MAT 02a]. The characteristics of the SPRITZ are as follows:

1) Frequency: 5.77 GHz CW, no modulation.

2) One hundred circular microstrip antennas with 0.75λ element spacing.

3) Right-handed circular polarization.

4) DC power supply from solar cells (15% efficiency) approximately 166 W.

5) System: one HPA – feeder network (power divider of 1– 100) – 3-bit phase shifter in each antenna.

6) Microwave radiation: >25W.

7) Total efficiency: > 15%.

8) Spurious: <–77.5 dBc.

9) For demonstration, light-emitting diode (LED) should be lighted at distance of 1–2 m.

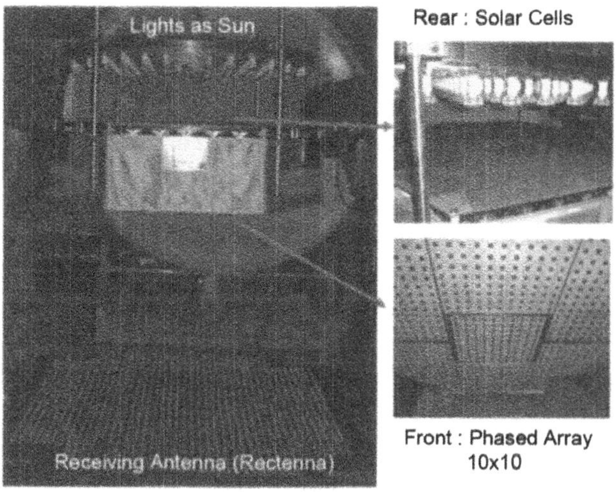

Figure 3.40. *SPRITZ in FY2000 – phased array for the SPS*

Total system efficiency of the phased array was 15%, which is without solar cells and that includes losses in the feeder network and a phase shifter. The efficiency was not enough; however, this was the first trial of the sandwich-type phased array, which was composed of rear solar cells and a front phased array. Theoretical and measured beam patterns are shown in Figure 3.41. The beam width was 7.3°. The antenna gain was 17.6 dBi and the equivalent isotropic radiated power (EIRP) was 58 dBm. A rectenna with 51.2%

efficiency received the microwave power and lighted an LED to detect the microwave beam.

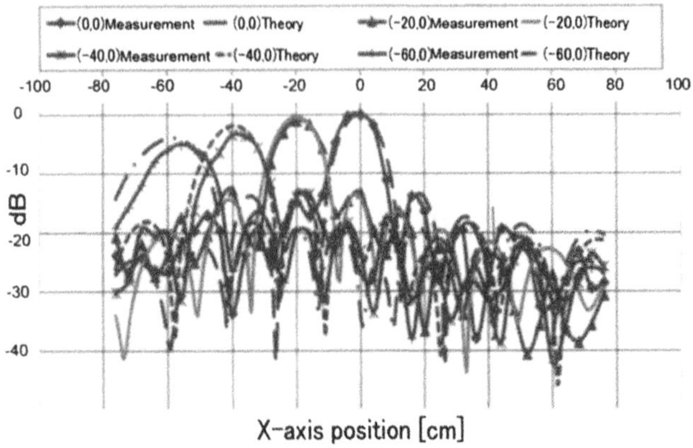

Figure 3.41. *Theoretical and measured beam pattern of SPRITZ phased array*

High efficiency is important in the phased array for MPT. Some new ideas to increase the efficiency or to decrease the loss against the development of high-efficiency microwave circuits were required. In FY2001, Kyoto University and Mistubishi Electric Corporation developed a parabolic antenna phased array at 5.8 GHz to retain high efficiency of the microwave circuits (Figure 3.42). The loss is evident in all amplifiers (except in microwave HPAs), for example driver amplifiers, phase shifter and beam control circuits, isolators or occurrence of grating lobes. Therefore, to decrease the power loss, a parabolic antenna phased array with the following ideas was developed [IKE 02, MIK 04].

1) Direct digital synthesizer (DDS)/PLL oscillators to reduce phase shifters.

2) Phased array radiator to control the element factor to suppress grating lobes.

3) Parabolic antenna phased array to concentrate the microwave power (high gain).

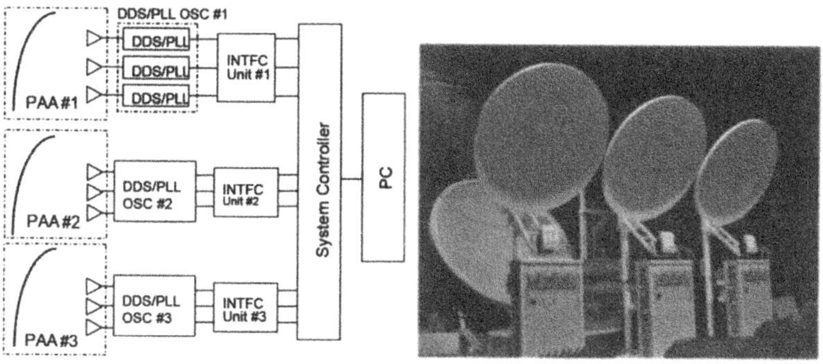

a) b)

Figure 3.42. *Parabolic phased array with DDS / PLL oscillators. a) System block diagram; b) image*

One DDS/PLL oscillator generates more than 8 W of continuous microwave power. Three DDS/PLLs are used in one parabolic antenna and the phased array is composed of three parabolic antennas, thus the total radiated power is more than 72 W (8 W × 3 W × 3 W). Diameter and gain of the reflector are 1.2 m and 32.2 dB, respectively. The controllable beam direction is within 5° on the horizontal line. Parabolic antenna spacing is 1.25 m. Theoretical and measured beam patterns at the Fresnel region (7 m away from the baseline) are shown in Figure 3.43. We can control the beam direction by suppressing grating lobes with the parabolic antenna phased array.

USEF SPS Study Team in Japan developed some types of phased array for SPS. In FY2002, the USEF SPS Study Team and Mitsubishi Heavy Industry developed a phased array whose characteristics were as follows [KIM 04]:

1) Frequency: 5.77 GHz CW, no modulation.

2) Nine circular microstrip antennas with 0.75λ element spacing.

3) HPA efficiency and power: >2 W and >51% (Figure 3.44).

4) HPA: class-AB amplifier.

5) Four-bit digital phase shifters.

6) Spurious: <–50 dBc at second and third harmonics.

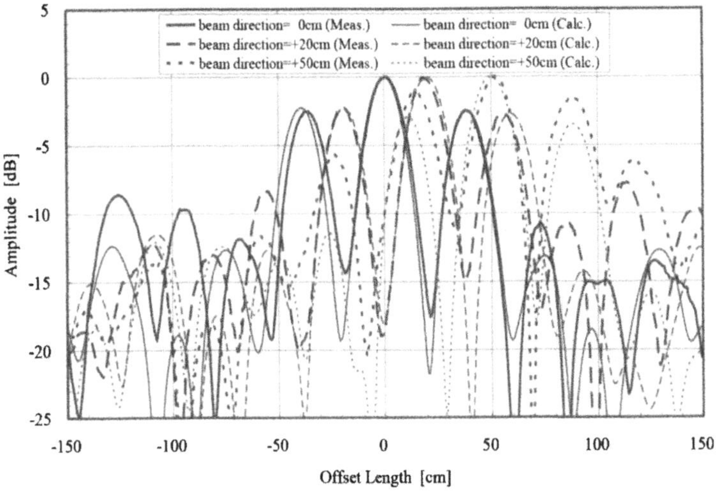

Figure 3.43. *Theoretical and measured beam pattern of parabolic antenna phased array at Fresnel region*

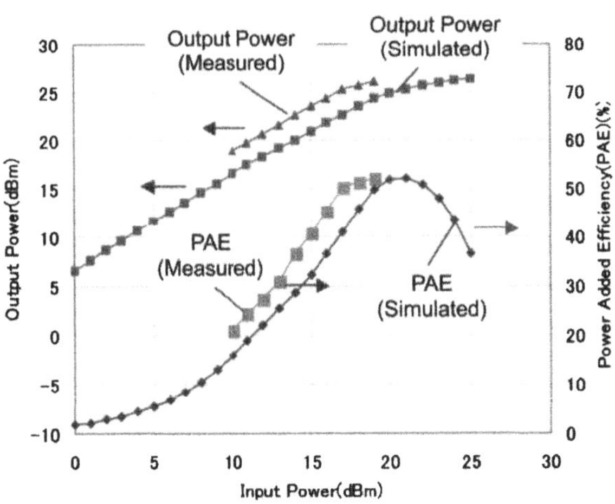

Figure 3.44. *Measured output power, efficiency and gain of the GaAs amplifiers in AIA developed by USEF SPS Study Team*

The phased array was composed of three layers with receiving, phase-shifting, and transmitting parts as an AIA (Figure 3.45). The microwave source was fed from the backside of the AIA through the air and the received microwave was amplified and controlled through the phase-shifting and transmitting parts. The array was approximately 360 mm × 360 mm × 70 mm and the antenna gain was 10.8 dBi. The total weight was approximately 11 kg. The radiation efficiency of transmission antennas was more than 90% in approximately 37° range of beam width (Figure 4.46). The measurement result of the beam pattern is shown in Figure 4.47, when the microwave beam was controlled in the horizontal directions.

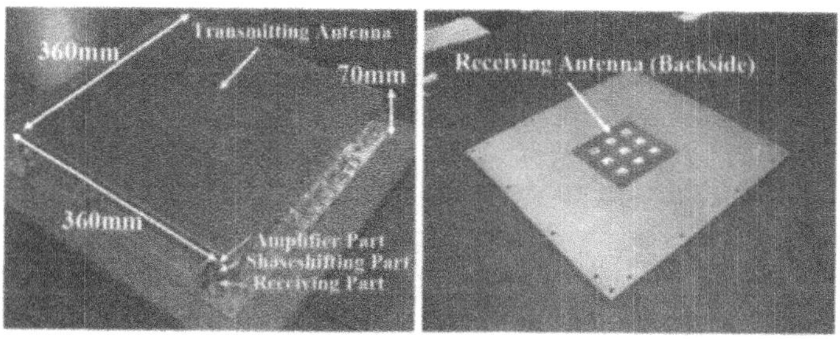

a) b)

Figure 3.45. *Active integrated antenna for MPT in FY2002. a) Front; b) back*

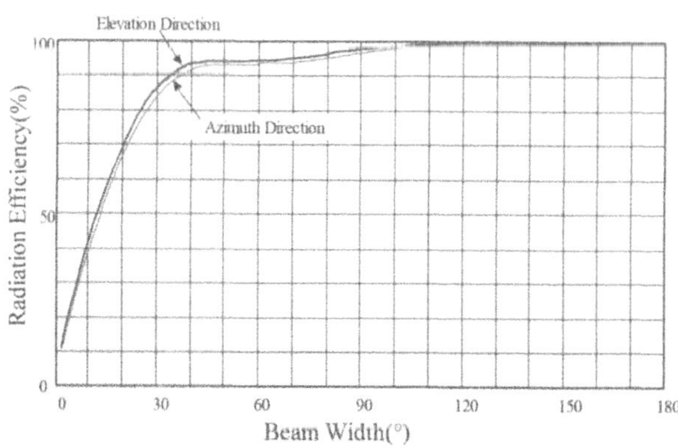

Figure 3.46. *Radiation efficiency versus beam width*

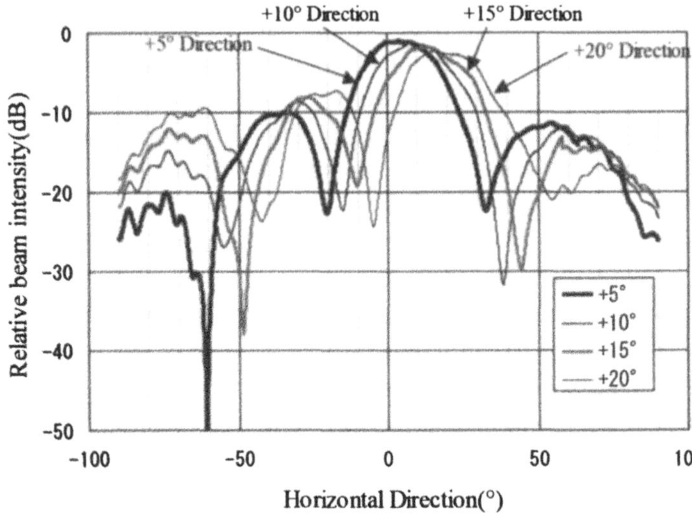

Figure 3.47. *Measured beam pattern of USEF-AIA*

USEF continued to develop a high-efficiency phased array for SPS. The basic plan for space policy was established by the Strategic Headquarters for Space Policy in June 2009. This basic plan for space policy that was forged at this time was based on the Basic Space Law established in May 2008 and is the Japan's first basic policy relating to space activities. In the plan, SPS was selected based on nine systems and programs for the use and R&D of space [BAS 08a, BAS 08b].

Based on this, a high-efficiency and thin phased array development project, which is supported by METI, started in FY2009 [FUS 11]. The author is a chair person of this project and Mitsubishi Electric Corporation is developing the phased array. The target of the phased array is: (1) 5.8 GHz CW, no modulation, (2) >70% PAE GaN semiconductor MMIC amplifier (Figure 3.48), (3) MMIC 5-bit phase shifters, (4) <40 mm thick phased array (Figure 3.49), (5) 120 cm × 120 cm array, (6) 76 amplifier/phase shifter modules on each panel of a four-panel system (76 × 4 = 304 modules in system), (7) four antenna – one module subarray and (8) total power > 1.6 kW CW.

Figure 3.48. *Metal packaged GaN HEMT amplifier [HOM 11]*

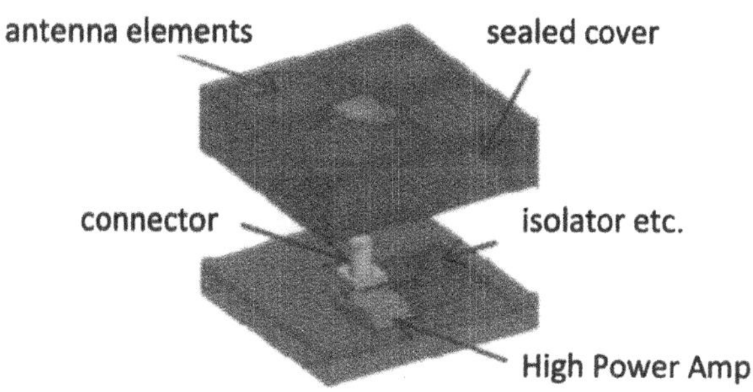

Figure 3.49. *Structure image of sub-array [HOM 11]*

At the end of FY2010, a new phased array was installed in Kyoto University as multipurpose research equipment (Figure 3.50) [HOM 11, YAM 10]. The characteristics of the phased array in Kyoto University are as follows:

1) Frequency: 5.8 GHz CW, no modulation.

2) Separated module antenna/active circuits system.

3) Rigid antenna plane.

4) A total of 256 elements.

5) Active phased array with one active circuit for one antenna.

6) Output microwave power of 1.5 kW.

7) Class F power amplifiers with GaN FETs.

8) Output >7 W in HPA as final stage;

9) PAE >70% in microwave HPA as the final stage (Figure 3.51).

10) Total DC-microwave conversion efficiency >40%.

11) Five-bit MMIC phase shifters.

12) Thickness <30 cm as universal experimental equipment.

Figure 3.50. *New phased array for MPT with GaN FET and class F amplifier circuits developed in Kyoto University and by Mitsubishi Electric Cooperation in FY2010*

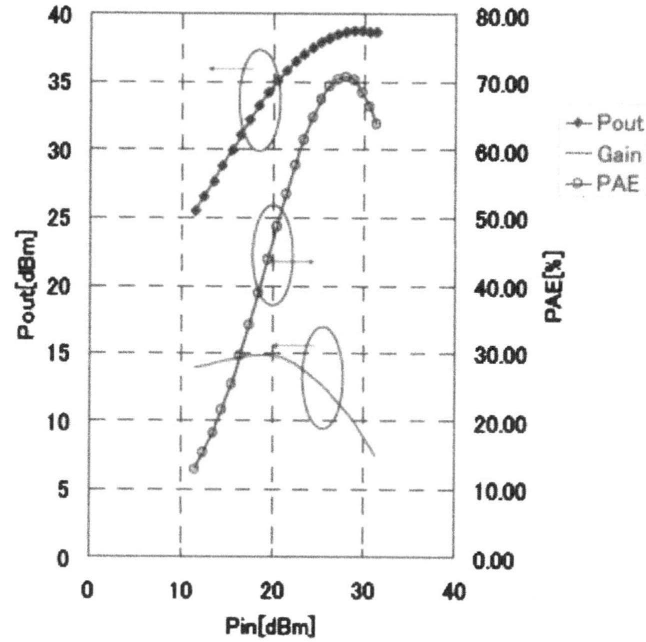

Figure 3.51. *Measured output power, efficiency and gain of the GaN amplifiers in phased array in Kyoto University*

The phased array system consists of phased array equipment, beam control units and a cooling unit. The beam control units consist of an antenna control unit, a PC and the retrodirective equipment. The rectenna array system consists of the rectenna array, a DC/DC converter, a load and the retrodirective equipment. Figure 3.52 shows a simulated beam pattern, and Figure 3.53 shows measured beam patterns, when the main beam is steered to EL = –15°, -10°, –5°, 0°, 5°,10°, 15°. In each beam steering angle, the obtained steering accuracy was within 0.1°. The phased array is open for use by inter-universities and international collaborative studies.

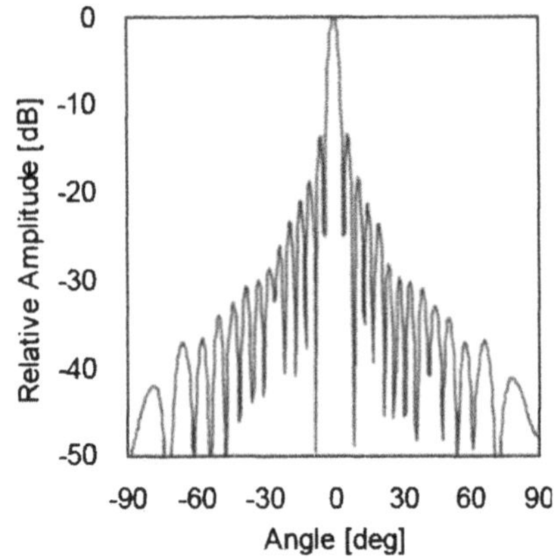

Figure 3.52. *Simulated beam pattern of the phased array in Kyoto University*

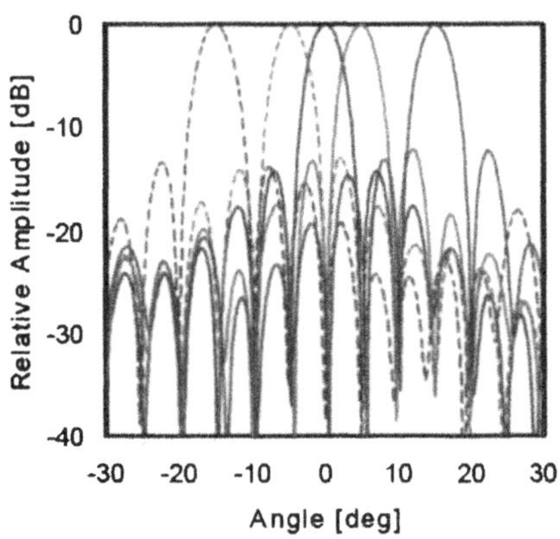

Figure 3.53. *Measured elevation beam pattern of phased array (beam steering angle: EL= –15°, –5°, 0°, 5°, 15°(AZ = 0°))*

3.4.4. *Phased array using magnetrons*

The compatibility of high-efficiency beam forming of MPT with semiconductor technology has not yet been realized. We at Kyoto University proposed and developed a new phased array with magnetrons. The efficiency of the magnetron is more than 70%, and it is the cheapest available microwave device. We cannot control the phase of the magnetron itself because it is a high-power generator. However, we developed a PCM with an injection-locking technique and a PLL feedback to a voltage source of the magnetron [SHI 03]. The original idea of PCM was given by Brown [BRO 88]. We revised the PCM and developed a phased array with PCM at 2.45 GHz in FY2000 and at 5.8 GHz in FY2001, which are called the Space Power Radio Transmission System for 2.45 GHz (SPORTS-2.45) and Space Power Radio Transmission System for 5.8 GHz (SPORTS-5.8), respectively [SHI 04b].

Characteristics of the SPORTS-2.45 are as follows:

1) Frequency: 2.45 GHz CW, no modulation.

2) Twelve PCMs.

3) Output power (matched load) of one PCM: >340 W

4) Efficiency of the PCM: >70.5%.

5) Total (12 PCMs) microwave power: >4 kW.

6) Five-bit digital phase shifters on each PCM.

7) Two types of array antenna (horns and dipoles as follows).

8) Retrodirective system with a CW pilot signal of 400 MHz.

In SPORTS-2.45, we always turn off the filament current during power transmitting after stable oscillation occurs because filament current causes noise in the magnetron. We cannot turn off the filament current in the cooker-type magnetron because a half-wave rectification voltage source is used for the microwave oven. The current heats the filament and the heat from the filament current usually supports electron emission to generate a microwave. Thus, enough electrons cannot be provided in the cooker-type magnetron with a half-wave rectification voltage source without any filament current. However, when we use a stabilized DC current source with the same cooker-type magnetron, the filament is heated enough without the filament current because the stabilized DC current obstructs the cooling of the filament. We achieved better than 10^{-8} frequency stability, relative to the frequency stability of an input reference signal with the stabilized DC power source and without the filament current.

We can select from two types of antenna in SPORTS-2.45 (Figure 3.54). One is a 12-horn antenna array with low power loss but with a limited narrow beam scanning capability. The size and gain of each horn antenna are 192 mm × 142 mm and 17.73 dBi, respectively. System efficiency of the horn array is high, but large side lobes and grating lobes appear when we change the beam direction. The simulated beam pattern in front direction is shown in Figure 3.55. When we change the beam direction to wide range, grating lobes arise.

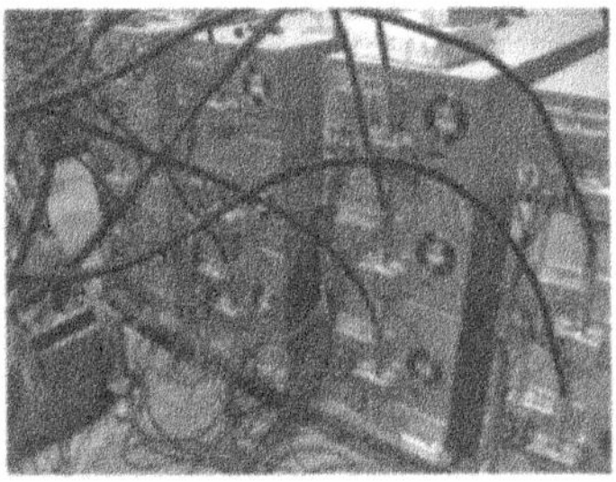

a)

b)

c)

Figure 3.54. *SPORTS-2.45 magnetron phased array; a) 12 phase controlled magnetrons, b) horn antenna array that is directly connected to the PCMs, c) antennas connected to PCMs through 1-bit subphase shifters after eight-way power dividers*

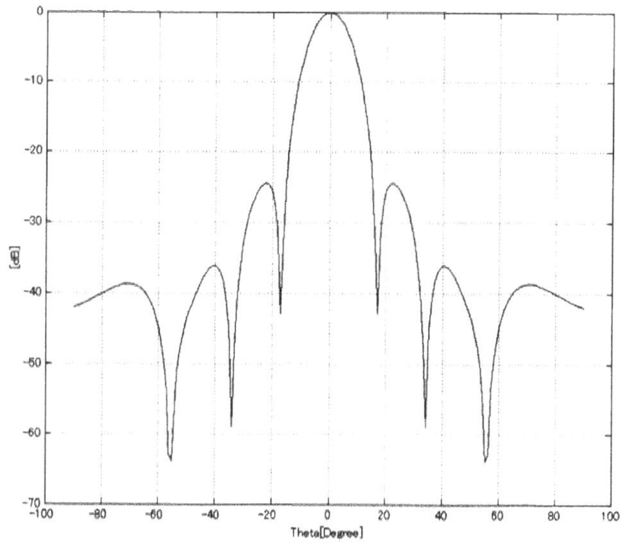

Figure 3.55. *Simulated beam pattern of 12-horn antenna array*

The other is a 96-dipole antenna array with power dividers and 1-bit subphase shifters. Element spacing is 0.7λ. To expand the beam control area without large side lobes and grating lobes, we should shorten element spacing. But the power of the PCM is too large to connect a small dipole. Thus, we should divide the power from the PCM to provide the small dipole. This is the concept of a subarray. The normal subarray makes a grating lobe that diffuses the microwave power except for the target when we change the beam direction. Therefore, we proposed a 1-bit subphase shifter that is installed after a power divider to suppress the grating lobe. Loss in the digital phase shifter is commonly estimated as 1 dB/1 bit. Thus, the loss of the 1-bit subphase shifter is smaller than that of the 4- or 5-bit phase shifters. If we do not use any subphase shifters, then the loss is the same as a subarray system and large side lobes and grating lobes appear. However, we can suppress the grating lobes with the subphase shifter system and can retain high efficiency of the phased array [SHI 04].

Characteristics of the SPORTS-5.8 are as follows:

1) Frequency: 5.77 GHz CW, no modulation.

2) Nine PCMs.

3) Output power (matched load) of one PCM: >300 W.

4) Efficiency of the PCM: >70%.

5) Total (nine PCMs) microwave power: >1.26 kW

6) Four-bit digital phase shifters on each PCM.

7) Total of 288 microstrip antenna in which one PCM is connected to 32 microstrip antenna elements each.

8) Retrodirective system with a CDMA modulated pilot signal of 4.8 GHz.

In SPORTS-5.8, we only adopt a subarray with power dividers after the 5.8 GHz PCMs without any subphase shifters (Figure 3.56). The power loss after output of PCM is less than 1.5 dB. Matsushita Co. developed the 5.8 GHz magnetron. We always turn off the filament current during power transmitting after the stable oscillation occurs.

Figure 3.56. *SPORTS-5.8 magnetron phased array*

In 2009, Kyoto University succeeded in a field MPT experiment using PCM technology. We transmitted 2.46 GHz microwave power with two 110 W output power PCMs from an airship to the ground [MIT 10]. Two radial slot antennas were used in the experiment, whose diameter was 72 cm, with a gain and aperture efficiency of 22.7 dBi and 54.6%, respectively. Element spacing was 116 cm (Figure 3.57). Theoretical and measured beam patterns are shown in Figures 3.58 and 3.59, respectively. Characteristics of this magnetron phased array are as follows:

1) Frequency: 2.46 GHz CW, no modulation.

2) Two PCMs.

3) Output power (matched load) of one PCM: >110 W.

4) Analog phase shifters on each PCM.

5) Two radio slot antennas.

6) Weight: <45 kg (transmitters, antenna, batteries, telemeters, etc.).

7) Retrodirective system with a CW pilot signal of 5.8 GHz.

8) Data links between the airship and ground with 2.45 GHz wireless LAN and 429 MHz specified low-power radio.

Figure 3.57. *Radial slot antenna array with two PCMs for MPT airship experiment in 2009 in Japan*

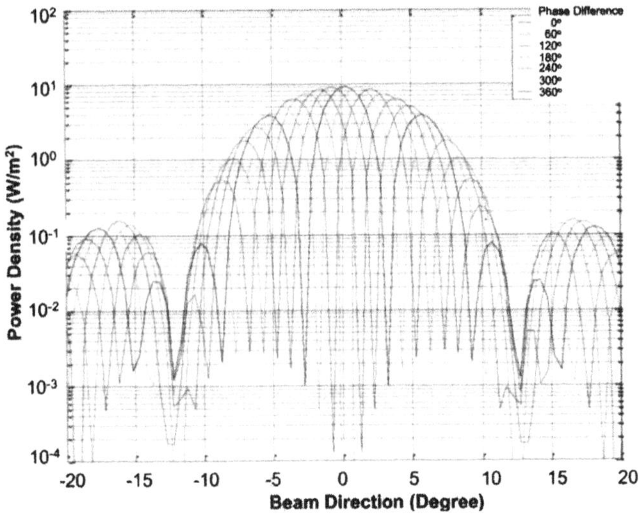

Figure 3.58. *Theoretical beam pattern of two radial slot antenna arrays. For a color version of this figure, see www.iste.co.uk/shinohara/radiowaves.zip*

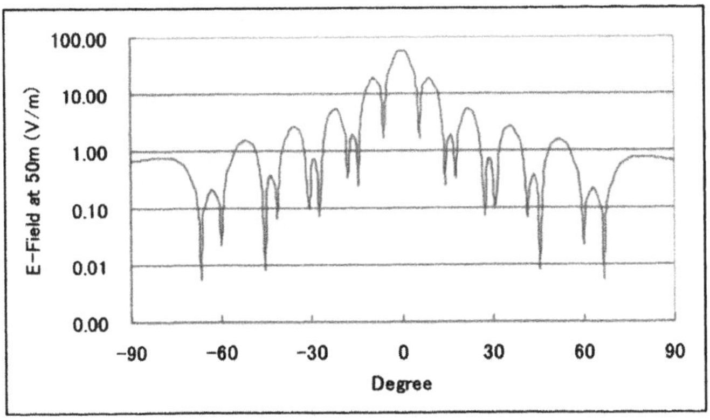

Figure 3.59. *Measured beam pattern (E-field) of two radial slot antenna arrays at 50 m distance. Radiated microwave power is 600 W*

In Kyoto University and Kyushu Technical University, we also apply the PCM phased array for MPT to a Mars observation airplane [NAG 11a]. We consider the magnetron phased array as one of the solutions for a high-efficiency phased array system.

3.4.5. *Retrodirective system*

Target detection is important for the phased array system because of accurate and high-efficiency microwave power transfer. Retrodirective target detection, in which a pilot signal is used both for the target detection and for detecting antenna positioning, is often used for the MPT system. An extensive research is being conducted on retrodirective array research for wireless communications in the world [MIY 02]. The developed retrodirective systems for various objectives are shown in Figure 3.60. [QUB, DID 98, QIA 98, MIY 01].

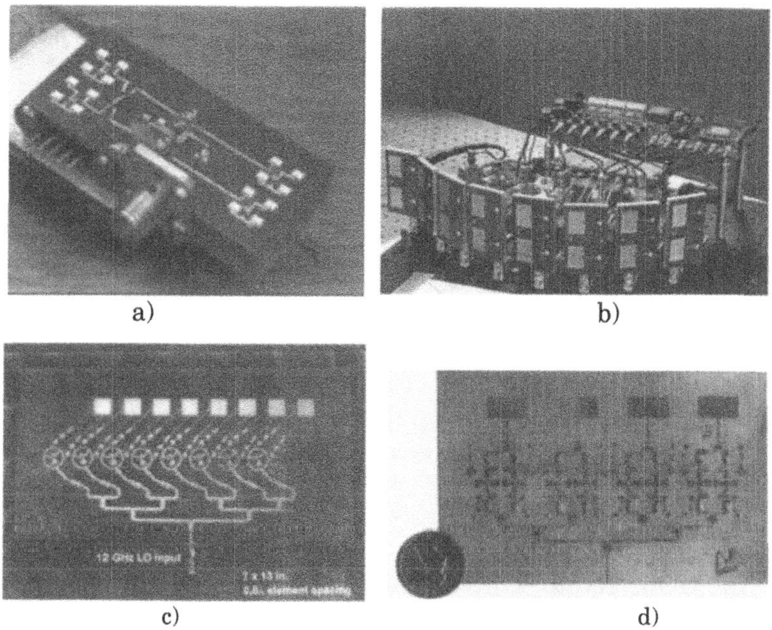

Figure 3.60. *Various retrodirective arrays with phase conjugate circuits developed in a) Queen's University (62–66 GHz)[HTT], b) Jet Propulsion Laboratory and University of Michigan in 2001 (5.9 GHz)[DID 98], c) UCLA in 1995 (6 GHz)[QIA 98], and d) UCLA in 2000 (6 GHz)[MIY 01]*

In Japan, some types of retrodirective system were developed after the 1980s. In 1987, Kyoto University and Mitsubishi Electric Corporation developed a retrodirective system with two asymmetric pilot signals (Figure 3.61) [MAT 02b]. We use the same frequency for the pilot signal and the transmitting microwave in the conventional retrodirective system. Therefore, interference between the pilot signal and the transmitting microwave may occur. Therefore, they proposed two asymmetric pilot signal systems with $\omega_t + \Delta\omega$ and $\omega_t + 2\Delta\omega$. Seven antennas were used for receiving two pilot signals as well as for microwave power transfer. The microwave frequency was 2.45 GHz.

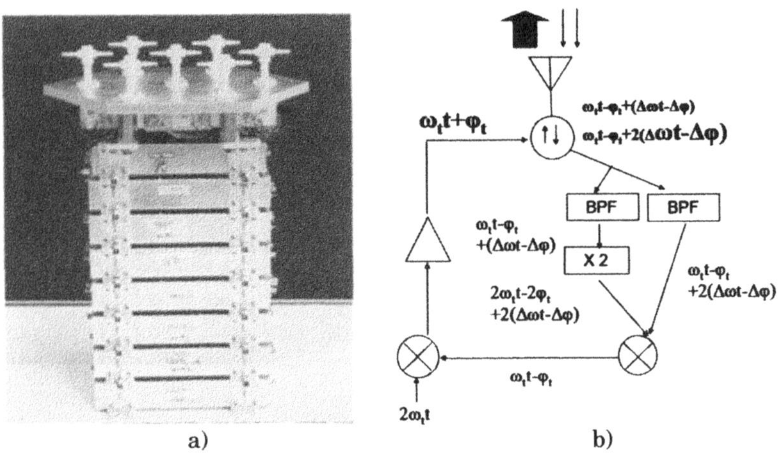

Figure 3.61. *Retrodirective system with two asymmetric pilot signals developed by Kyoto University and Mitsubishi Electric Corporation. a) Image of the phased array. b) Block diagram of retrodirective system with two asymmetric pilot signals*

In 1996, Kyoto University and Nissan Motor Company (presently IHI Aerospace) developed another retrodirective system. To suppress interference between the pilot signal and the transmitting microwave, they proposed to use a one-third frequency pilot signal (817 MHz) (Figure 3.62) [MAT 02a]. Transmitting microwave frequency was at 2.45 GHz. Eight transmitting antennas were put in one-dimensional line. On both sides of the transmitting antennas, pilot signal receiving antennas were placed. When we consider only a one-dimensional horizontal line, each pilot signal receiving antenna corresponds to a transmitting antenna. It did not contain a local oscillator to reduce unmatched frequency between a pilot signal and the local oscillator. Figure 3.63 shows the measured beam pattern of this retrodirective system. When there is no retrodirective target detected, the data indicate the array pattern itself. When the retrodirective system is on, the microwave beam chases the pilot signal source and the data indicate the element pattern.

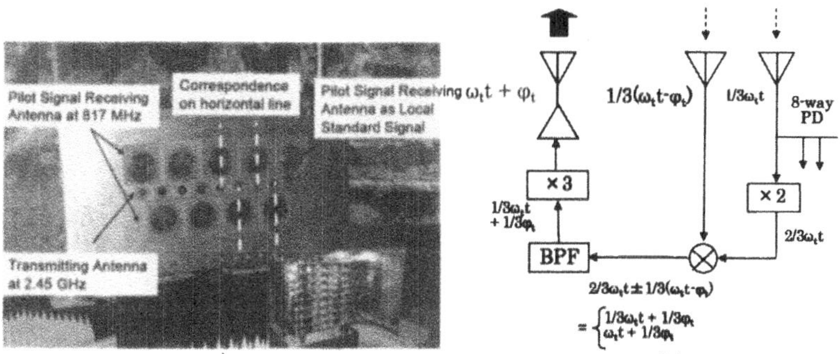

a) b)

Figure 3.62. *Retrodirective system with 1/3 frequency pilot signal developed by Kyoto University and Nissan Motor Company. a) Image of the phased array. b) Block diagram of retrodirective system with 1/3 frequency pilot signal*

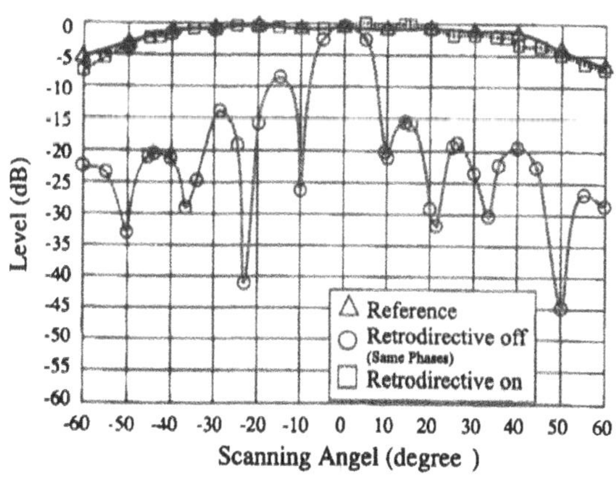

Figure 3.63. *Measured beam pattern of retrodirective system in on and off state*

In FY2003, the USEF SPS Study Team together with Mistubishi Electric Corporation developed a PLL-heterodyne retrodirective system [MIZ 04]. Microwave frequency was 5.77 GHz and the pilot signal frequency was 3.884 GHz. There were eight transmitting antennas (the other antennas in Figure 3.64 are dummies). Beam patterns are shown in Figures 3.65(a)–(c).

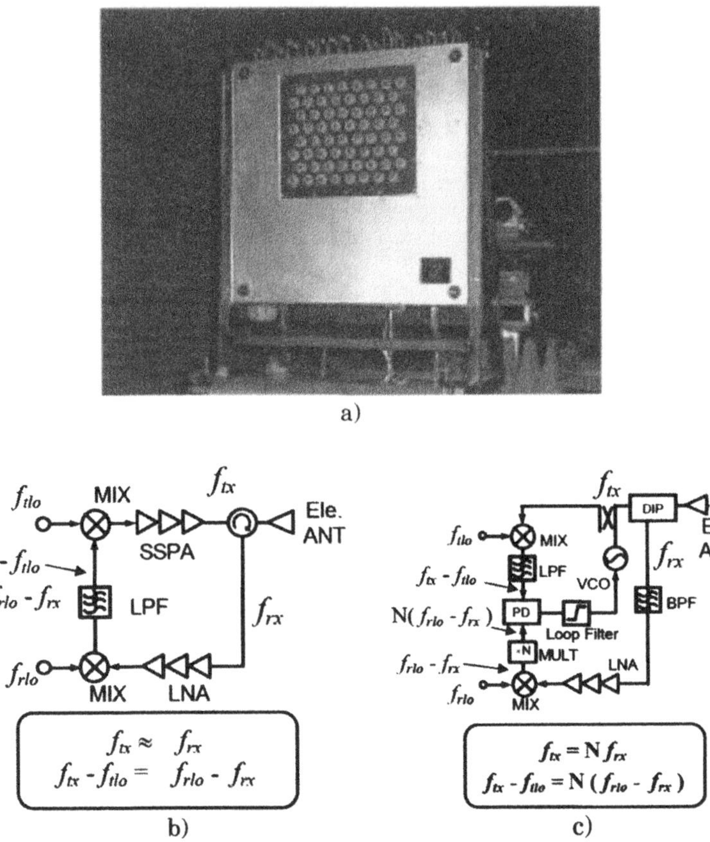

Figure 3.64. *PLL-heterodyne retrodirective system developed by Mitsubishi Electric Corporation with USEF SPS Study Team in FY2003 [MAT 02b]. a) Image and b) block diagram of conventional retrodirective. c) Block diagram of PLL-heterodyne retrodirective*

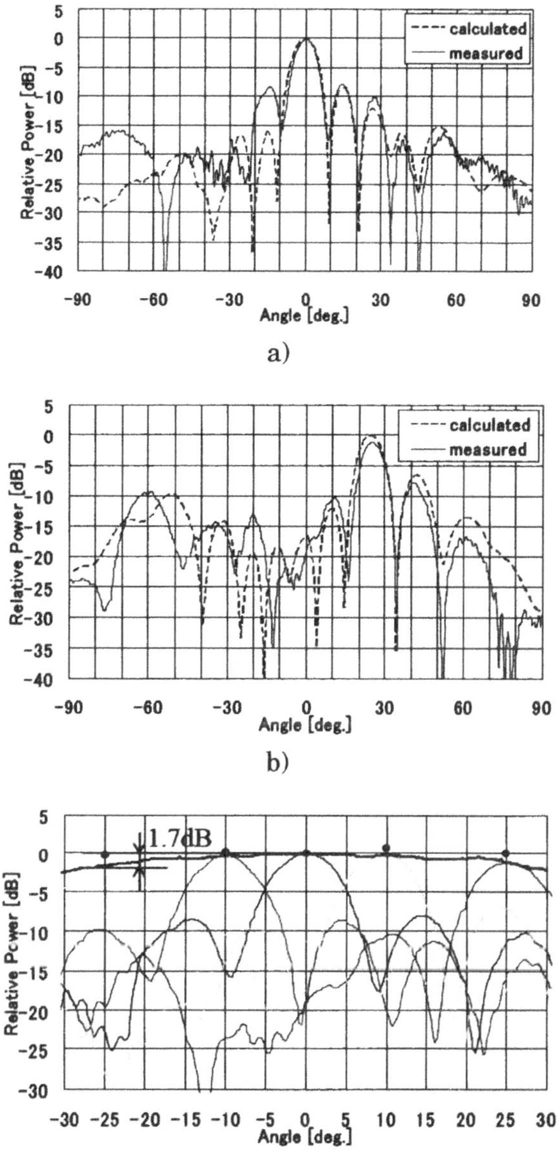

Figure 3.65. *Beam pattern of PLL-heterodyne retrodirective system [MAT 02b]. a) Beam target: 0°. b) Beam target: 25°. c) Beam pattern with and without retrodirective*

In January 2006, Kobe University, ISAS and ESA's research group succeeded in a retrodirective MPT experiment from a rocket to the ground [NAK 05, KAY 06]. They opened the antenna and structure on the rocket and transmitted microwaves to the pilot signal target.

3.5. RF rectifier – rectenna and tube type

3.5.1. *General rectifying theory of rectenna*

A rectenna is a passive element with rectifying diodes that operates without an internal power source. It can receive microwave power and rectify it to produce DC power. A general block diagram of a rectenna is shown in Figure 3.66. A low-pass filter is installed between the antenna and the rectifying circuit to suppress reradiation of higher harmonics from the diodes. An output filter is used not only to stabilize the DC current, but also to increase RF–DC conversion efficiency.

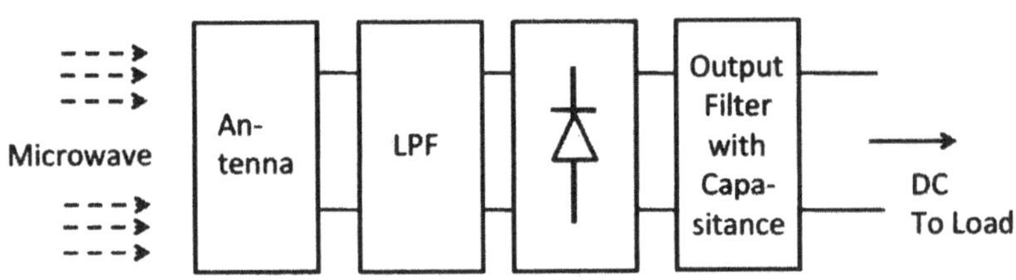

Figure 3.66. *General block diagram of a rectenna*

We can apply various antennas and rectifying circuits. Their selection depends on the requirements for a wireless power system and its users. When we use a rectenna array, its antennas are capable of absorbing 100% of the input microwaves [DIA 68, STA 74, ITO 86]. Higher efficiency rectifying circuits are required because a WPT system is an energy system. There are various rectifying circuits that can theoretically reach 100% efficiency. The details of general rectifying circuits are described in [SAE 11] and [OHI 13].

In MPT, a single-shunt full-wave rectifier is often used. The single-shunt rectifier is composed of a diode and a capacitor connected in parallel to a $\lambda_g/4$ distributed line. λ_g is the effective wavelength of an input radiowave (microwave or millimeter wave). The $\lambda_g/4$ distributed line and the parallel capacitance work in tandem as an output filter. The single-shunt rectifier can theoretically rectify microwave input at 100% efficiency with only one diode because of the enhancing effect of the output filter [GUT 79].

Figure 3.67 shows the block diagram of a rectenna with a single-shunt rectifier and its theoretical output waveform along with its analysis. A low-pass filter and a capacitance for DC conversion are usually installed between an antenna and a rectifier. An input impedance Z to the right of the diode is formulated as follows:

$$Z_L = \frac{R_L}{1 + j\omega C R_L} \qquad [3.8]$$

$$Z = Z_0 \frac{Z_L + jZ_0 \tan(\beta l)}{Z_0 + jZ_L \tan(\beta l)} \qquad [3.9]$$

where R_L is the load resistance, C is the capacitance of the output filter, Z_0 is the characteristic impedance of a distributed line and its circuit, l is the length of the distributed line, β is a phase constant ($\beta = 2\pi\lambda_g$) and ω is the angular frequency of the input electromagnetic wave. For odd harmonics, 3ω, 5ω, ..., $\tan\beta l$ and Z become infinity; when $l = \lambda_g/4$ and C is large enough, then $Z = Z_0^2/Z_L$. For even harmonics, 2ω, 4ω, ..., $\tan\beta l$ and Z become zero; when $l = \lambda_g/4$ and C is large enough, then $Z = Z_L$. These are the most important concepts regarding the theory of the single-shunt rectifier [GUT 79].

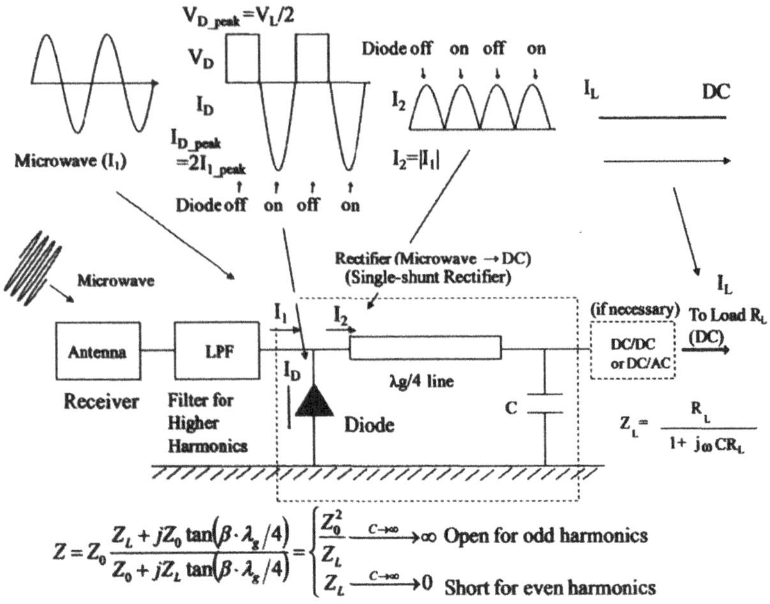

Figure 3.67. *Block diagram of a rectenna with a single-shunt full-wave rectifier and its theoretical waveform*

$I_1(t)$ is the input current through the LPF, defined as follows:

$$I_1(t) = I_0 \sin(\omega t) \tag{3.10}$$

where t is the time and I_0 is the amplitude of the current of an input electromagnetic wave. In WPT theory, we usually assume a pure, continuous and unmodulated wave.

When the diode is open at $0 < \omega t < \pi$, I_2 flows to the right-hand side of the diode, and I_D is the diode current. These currents are defined as follows:

$$I_2(t) = I_0 \sin(\omega t) \qquad (0 < \omega t < \pi) \tag{3.11}$$

$$I_D(t) = 0 \qquad (0 < \omega t < \pi) \tag{3.12}$$

As previously discussed, I_2 must be composed of even harmonics to obtain the efficiency enhancement effect of the output filter. As a result, I_2 must be a function with a cycle of π described as follows:

$$I_2(\omega t+\pi)=I_2(\omega t) \quad [3.14]$$

When the diode is short at π < ωt < 2π, then I_2 and I_D are:

$$I_2(t)=-I_0\sin(\omega t) \qquad (\pi<\omega t<2\pi) \quad [3.15]$$

$$I_D(t)=2I_0\sin(\omega t) \qquad (\pi<\omega t<2\pi) \quad [3.16]$$

As a result, I_2 becomes a full-wave waveform described as follows:

$$I_2(t)=I_0\left|\sin(\omega t)\right| \qquad (0<\omega t<2\pi) \quad [3.17]$$

$$I_D(t)=I_0\left|\sin(\omega t)\right|-I_0\sin(\omega t) \qquad (0<\omega t<2\pi) \quad [3.18]$$

In the diode, the waveform becomes a half-wave doubler. Thus, the behavior of the single-shunt rectifier is very similar to that of a class F amplifier.

The class F rectenna, which is not the single-shunt rectenna, was originally proposed by Kyoto University (Figure 3.68) [HAT 13]. A second group developed a class F rectenna with 77.9% efficiency at 2.45 GHz [NOD 12b]; theoretical analysis of the class F rectenna is described in [GUO 13]. Class C, class E and class inverse-F rectennas have also been proposed and developed [ROB 12, RUI 12]. In these reports, the RF–DC conversion efficiencies of class C, class E and class inverse-F rectennas were 72.8% at 2.45 GHz, 83% at 950 MHz with HEMT diodes and 85% at 2.14 GHz with GaN HEMT diodes, respectively.

Many rectennas using a single-shunt rectifier have demonstrated 70–80% RF–DC conversion efficiency [SHI 12]. Compared to a four-diode bridge rectifier, the efficiency of the single-shunt rectifier is higher because there is only one diode with a resistance loss factor in the single-shunt rectifier.

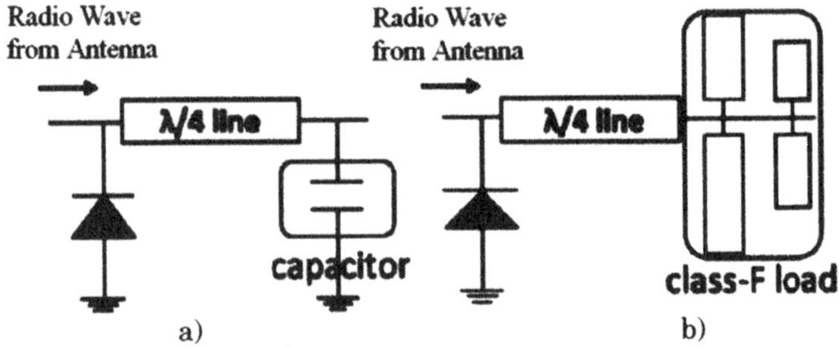

Figure 3.68. *Concept of class F load rectenna; a) normal single shunt, b) proposed class F load type*

Theoretically, the RF–DC conversion efficiency of a rectenna is 100%. But a real diode and a real circuit introduce loss factors. The V-I characteristics of the diode, in particular, determine the RF–DC conversion efficiency [YOO 92, MCS 98]. Figure 3.69 shows typical RF–DC conversion efficiencies of rectennas, showing not only single shunt, but also all rectifying circuits with diodes. V_J is the junction voltage of a diode and V_{br} is the breakdown voltage of a diode, and their characteristic curves are shown in Figure 3.70. These characteristics also hold for a connected load instead of the input power.

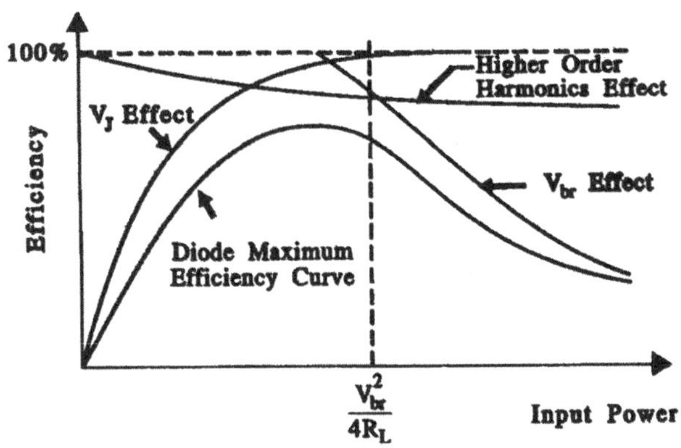

Figure 3.69. *Typical relationships between RF–DC conversion efficiency and input power [YOO 92]*

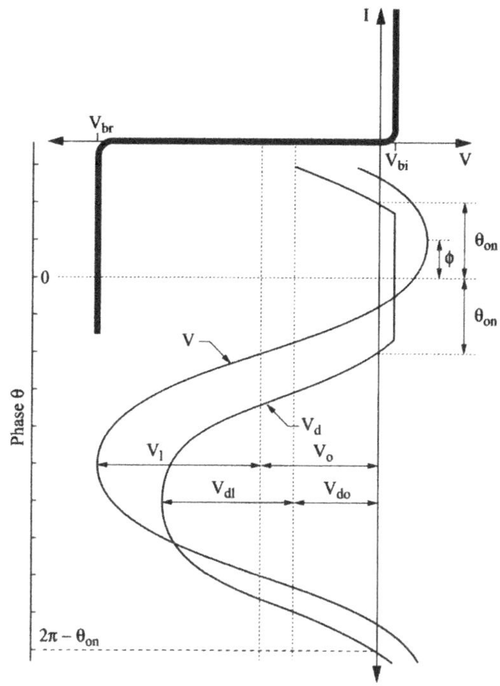

Figure 3.70. *Rectification cycle represented by an input fundamental waveform and diode junction voltage waveforms impressed on the diode V-I curve [MAC 98]*

The waveform at a diode can be expressed as:

$$V = -V_o + V_1 \cos\omega$$
$$V_d = \begin{cases} -V_{d0} + V_{d1}\cos(\omega t - \phi) & \text{if diode is off} \\ V_{bi} & \text{if diode is on} \end{cases} \qquad [3.19]$$

where V_o is the output self-bias DC voltage across the load and V_1 is the peak voltage amplitude of the input microwave. V_{d0} and V_{d1} are the DC and fundamental frequency components of the diode junction voltage and V_{bi} is the diode's built-in voltage in the forward bias region. The diode model consists of a series resistance R_s, a nonlinear junction resistance R_j described by its DC I-V characteristics and a nonlinear junction capacitance C_j as shown in Figure 3.71 [YOO 92, MCS 98]. A DC load resistance R_L is connected in parallel to the diode along a DC path represented by the dotted line to complete the DC circuit. The junction resistance is assumed to be zero for forward bias and infinite for reverse bias. By applying Kirchoff's voltage law, closed-form equations for the diode's efficiency and input impedance are determined. The RF–DC conversion efficiency η . is theoretically calculated from equation [3.20] [YOO 92, MCS 98]:

$$\eta = \frac{1}{1+A+B+C}$$

$$A = \frac{R_L}{\pi R_S}\left(1+\frac{V_{bi}}{V_o}\right)^2\left[\theta_{on}\left(1+\frac{1}{2\cos^2\theta_{on}}\right)-\frac{3}{2}\tan\theta_{on}\right]$$

$$B = \frac{R_S R_L C_j^2 \omega^2}{2\pi}\left(1+\frac{V_{bi}}{V_o}\right)\left(\frac{\pi-\theta_{on}}{\cos^2\theta_{on}}+\tan\theta_{on}\right) \qquad [3.20]$$

$$C = \frac{R_L}{\pi R_S}\left(1+\frac{V_{bi}}{V_o}\right)\frac{V_{bi}}{V_o}(\tan\theta_{on}-\theta_{on})$$

θ_{on} is defined as follows:

$$\tan\theta_{on}-\theta_{on} = \frac{\pi R_S}{R_L\left(1+\frac{V_{bi}}{V_o}\right)} \qquad [3.21]$$

C_j is a nonlinear junction capacitance and is described with zero bias junction capacitance C_{j0} as follows:

$$C_j = C_{j0}\sqrt{\frac{V_{bi}}{V_{bi}+|V_o|}} \qquad [3.22]$$

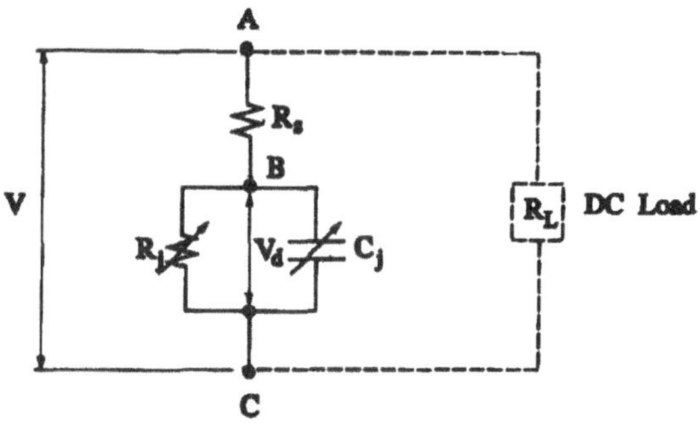

Figure 3.71. *Equivalent circuit of a diode [YOO 92]*

The estimated RF–DC conversion efficiency of a diode calculated using equations [3.20]–[3.22] for various circuit parameters is shown in Figure 3.72.

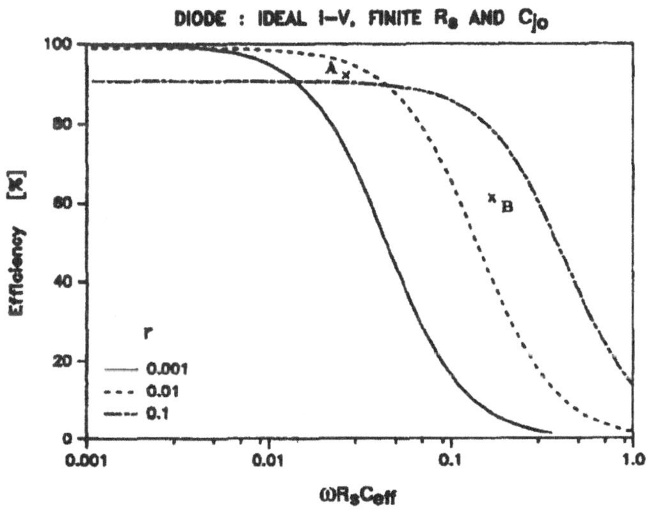

Figure 3.72. *RF–DC conversion efficiency of a diode estimated by frequency, internal resistance and capacitance ($r = R_s / R_L$) [MAC 98] (point A: R_s = 0.5 Ω, C_{j0} = 3 pf, 2.45 GHz, R_L = 100 Ω; point B: R_s = 4.85 Ω, C_{j0} = 0.13 pf, 35 GHz, R_L = 100 Ω) [YOO 92]*

3.5.2. *Various rectennas I – rectifying circuits*

The single shunt is not the only rectifying circuit available. Various rectennas have been proposed. The world's first rectenna developed by Brown in 1963 was a normal bridged rectifier (Figure 3.73) [BRO 80]. The operating frequency was 2.45 GHz. The frequency of 2.45 GHz (and 5.8 GHz) lies in the industrial, scientific and medical (ISM) band. The rectenna was a string-type rectenna developed for a fuel-free helicopter experiment in 1964. It was conceived at Raytheon Co., and a power output of 7 W was produced at approximately 40% efficiency with four point-contact diodes.

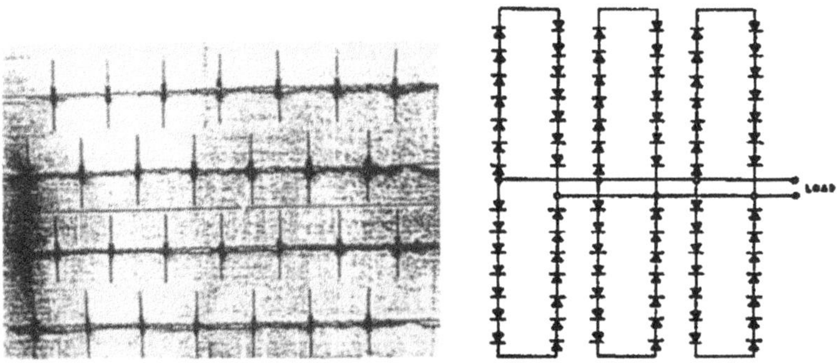

Figure 3.73. *World's first rectenna developed by Brown [BRO 80]*

Next, Brown developed the first single-shunt rectenna (Figure 3.74) [BRO 78]. It was developed for low-cost production. The same type of rectenna was adopted for a 1.6 mile MPT field experiment in Goldstone in 1975 [BRO 84]. Brown developed a large rectenna array with a size of 3.4 m × 7.2 m. The transmitted microwave power from the klystron source was 450 kW at a frequency of 2.388 GHz in the field experiment, and the achieved rectified DC power was 30 kW DC with 82.5% rectifying efficiency. At last, Brown achieved 90% efficiency of the rectenna at 2.45 GHz (Figure 3.75) [BRO 76]. In parallel, Brown also developed a thin rectenna shown in Figure 3.76 [BRO 82].

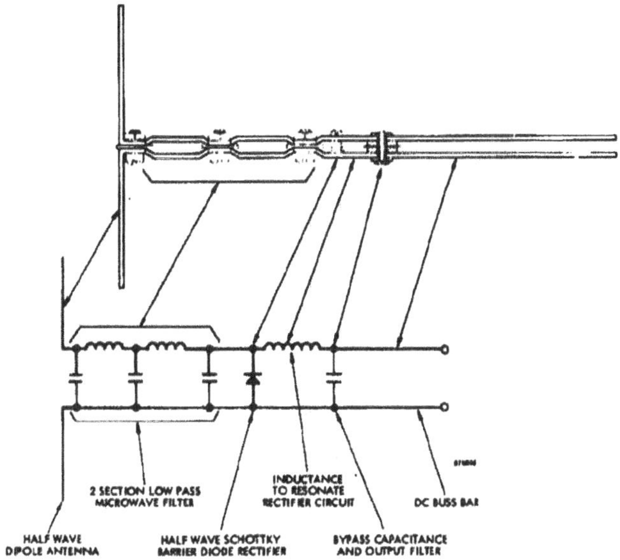

Figure 3.74. *Single-shunt rectenna developed by Brown [BRO 78]*

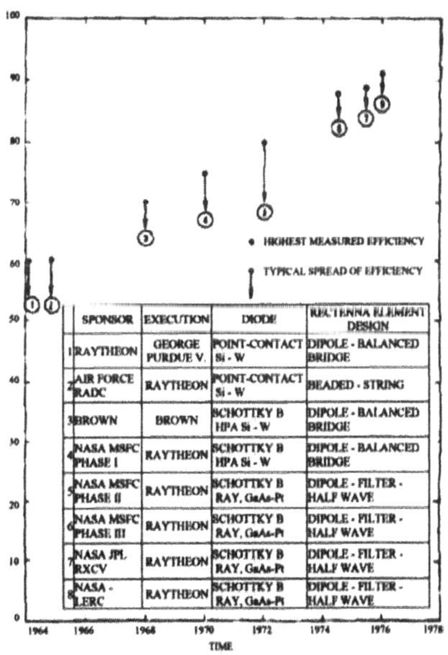

	SPONSOR	EXECUTION	DIODE	RECTENNA ELEMENT DESIGN
1	RAYTHEON	GEORGE PURDUE V.	POINT-CONTACT Si - W	DIPOLE - BALANCED BRIDGE
2	AIR FORCE RADC	RAYTHEON	POINT-CONTACT Si - W	BEADED - STRING
3	BROWN	BROWN	SCHOTTKY B HPA Si - W	DIPOLE - BALANCED BRIDGE
4	NASA MSFC PHASE I	RAYTHEON	SCHOTTKY B HPA Si - W	DIPOLE - BALANCED BRIDGE
5	NASA MSFC PHASE II	RAYTHEON	SCHOTTKY B RAY, GaAs-Pt	DIPOLE - FILTER - HALF WAVE
6	NASA MSFC PHASE III	RAYTHEON	SCHOTTKY B RAY, GaAs-Pt	DIPOLE - FILTER - HALF WAVE
7	NASA JPL RXCV	RAYTHEON	SCHOTTKY B RAY, GaAs-Pt	DIPOLE - FILTER - HALF WAVE
8	NASA - LERC	RAYTHEON	SCHOTTKY B RAY, GaAs-Pt	DIPOLE - FILTER - HALF WAVE

Figure 3.75. *Progress in optimized efficiency of rectenna by Brown [BRO 76]*

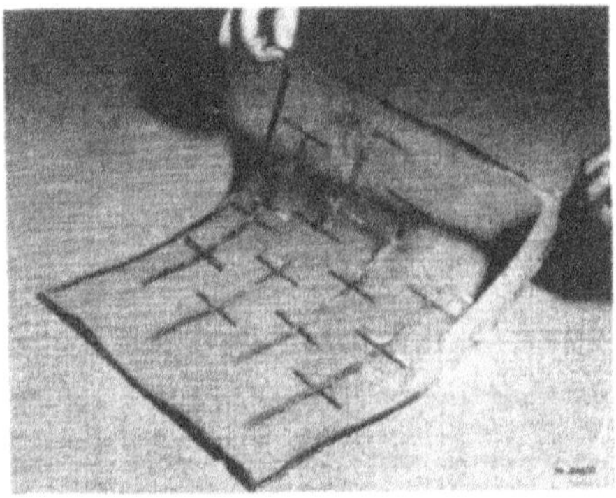

Figure 3.76. *Thin rectenna developed by Brown [BRO 82]*

After the progress of the rectenna achieved by Brown, some interesting approaches were considered to increase the efficiency. Figure 3.77 shows a full-wave rectifier with 0°–180° combination through an antenna [ITO 93]. In that rectifier, a combination of reverse phase microwave was used to realize full-wave rectifying. The antenna plays an important role in creating the reverse phase. The other approach using a 0°–180° combination had a rat-race hybrid circuit (Figure 3.78) [KOB 93]. The frequency was 2.45 GHz in both rectennas discussed in [ITO 93] and [KOB 93].

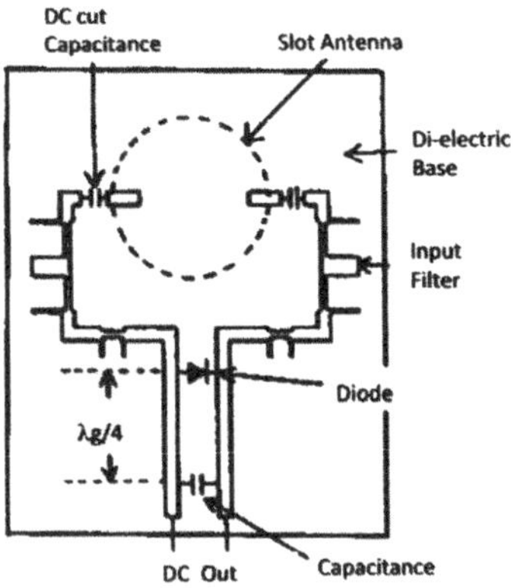

Figure 3.77. *Full-wave rectifier with 0°–180° combination through the antenna [ITO 93]*

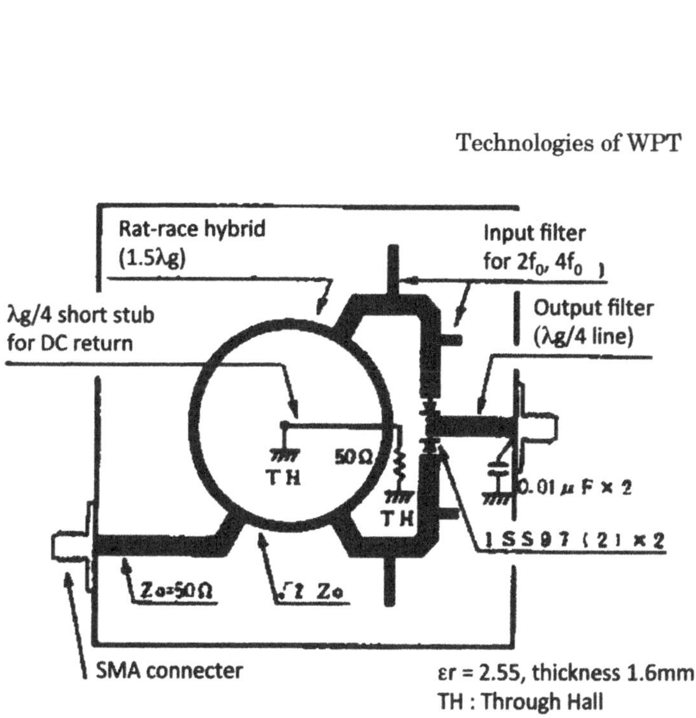

Figure 3.78. *Full-wave rectifier with 0°–180° combination through a rat-race hybrid (originally in Japanese) [KOB 93]*

A diode used in a rectifying circuit is a nonlinear device. Therefore, we cannot match its impedance to suppress reflection from the rectifying circuit with varying microwave input or connected load. When the optimum input microwave is input and the optimum load is connected to the rectifying circuit, the reflected microwave signal is less than 1%, but RF–DC the conversion efficiency becomes less than 10% and the reflected microwave signal is more than 50% when the input microwave and the connected load are not optimal (see Figure 3.69). Thus a rectenna with utilization of the reflected microwave was proposed by Kyoto University [KIM 02, KIM 03]. Three types of rectennas, called "feedback rectennas", were proposed and developed (Figure 3.79(a)), "feedback rectenna with rat-race hybrid" (Figure 3.79(b)) and "rectenna with direct utilization of reflected microwave" (Figure 3.79(c)), all operating at 2.45 GHz.

All rectennas in this chapter have diodes to rectify microwaves. But Professor Popovic's group adopted an FET to rectify microwaves. She proposed and developed a rectenna with FET, which is the same as a class inverse-F amplifier (Figure 3.80) [ROB 12]. Any amplifier will function as a rectifier, and at microwave frequencies as a self-synchronous rectifier without any gate drive Professor Popovic called this "PA-rectifier" and achieved 85% efficiency at 2.11 GHz, 8–10 W input. It can be used as a class inverse-F amplifier, whose PAE is 83% at 2.11 GHz, 8 W input.

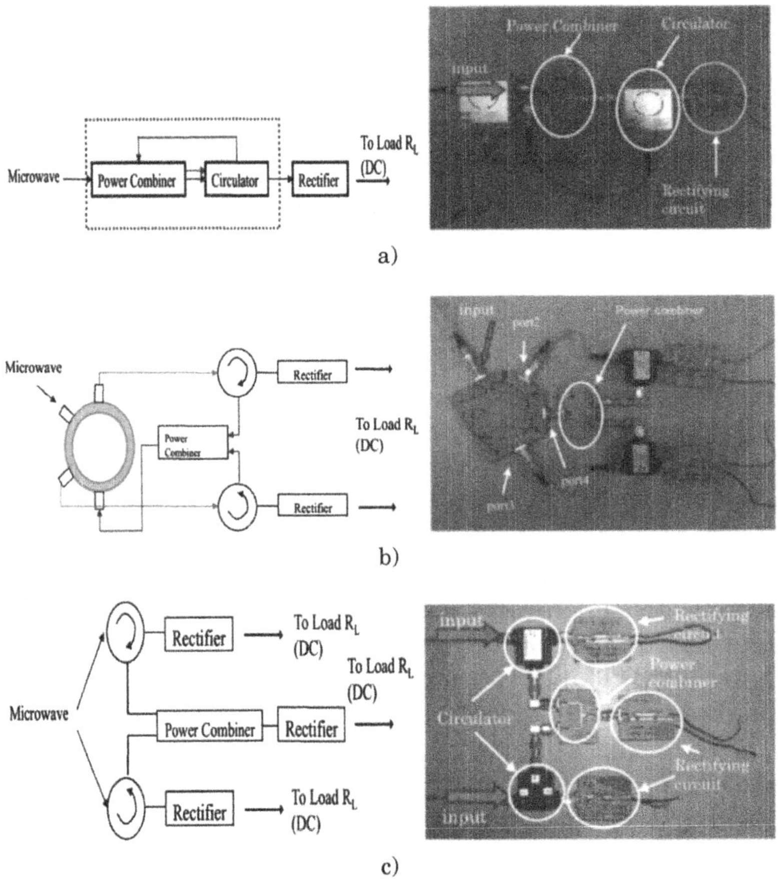

Figure 3.79. *a) Feedback rectenna, b) feedback rectenna with rat-race hybrid and c) rectenna with direct utilization of reflected microwave [KIM 02]*

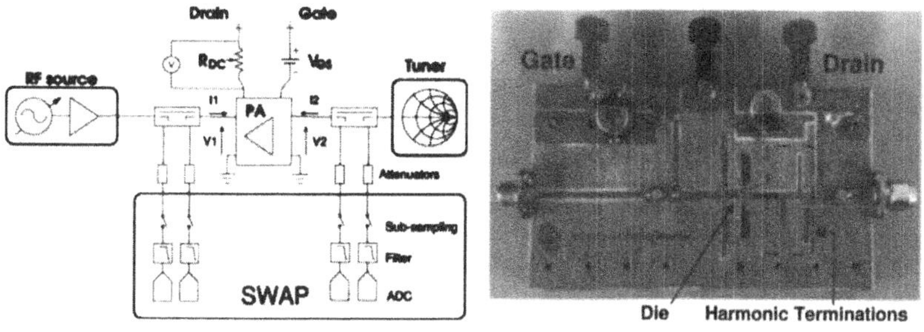

a) b)

Figure 3.80. *a) Block diagram of PA-rectifier, and b) the developed PA-rectifier with class inverse-F [ROB 12]*

3.5.3. *Various rectennas II – higher frequency and dual bands*

Antenna theory determines beam efficiency based on the parameters of antenna diameter, distance and frequency [SHI 11a]. Higher frequency will reduce the size of the antennas for MPT. Some trials were conducted for development of rectennas at higher frequencies.

The next ISM band after 2.45 GHz is 5.8 GHz. Research groups have developed 5.8 GHz rectennas (Figure 3.81) [MCS 98, SAK 97]. In particular, a rectenna developed at Texas A&M University has the highest RF–DC conversion efficiency at 5.8 GHz; the efficiency is more than 80% even when the load varies from 300 to 500 Ω with a voltage standing wave ratio (VSWR) of 1.29 at 50 mW of input microwave power [MCS 98].

Denso Corp. in Japan developed 14 GHz rectennas for a moving robot in a tube. The frequency was determined by the size of the tube. A monopole antenna and a voltage doubler were adopted in this system (Figure 3.82) [SHI 97]. RF–DC conversion efficiency was approximately 39% at 14–14.5 GHz, 100 mW and 2 kΩ.

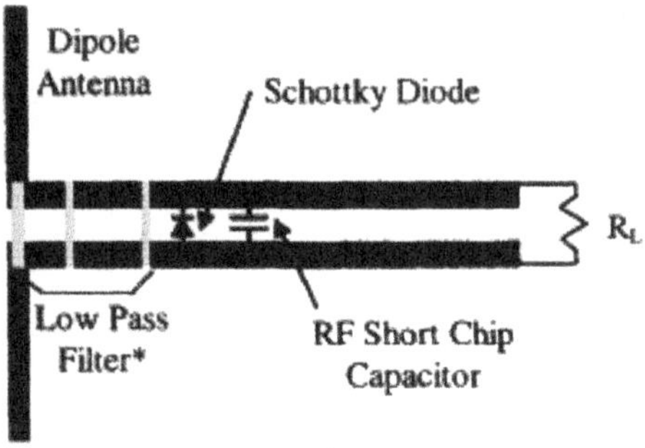

a)

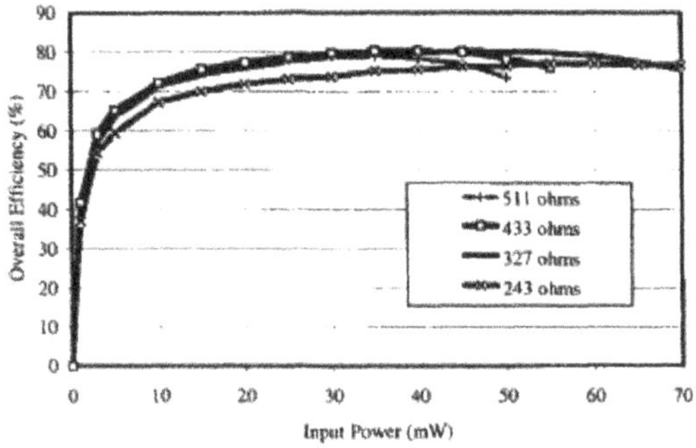

b)

Figure 3.81. *a) Rectenna with an antenna and single-shunt rectifier at 5.8 GHz and b) RF–DC conversion efficiency dependence of input 5.8 GHz microwave [MAC 98]*

a)

b)

Figure 3.82. *a) Rectifying circuit at 14 GHz and b) tube robot with a rectenna using a monopole antenna [SHB 97]*

Kyoto University and NTT Corp. in Japan collaborated to produce an MPT to feed a fixed wireless access (FWA) network. As a final example, we consider a simultaneous wireless system with information and power at millimeter wave levels. First, we developed a 24 GHz MMIC rectenna [HAT 13]. The frequency of 24 GHz is also on ISM band. We chose a class F load as an output filter to increase efficiency at higher frequencies. Dimensions of the 24 GHz MMIC rectenna are 1 mm × 3 mm on GaAs (Figure 3.83), with a maximum RF–DC conversion efficiency of 47.9% for a 210 mW microwave input signal at 24 GHz with a 120 Ω load. A Canadian group and a Spanish group have also developed 24 GHz rectennas for energy harvesting [LAD 13, COL 13].

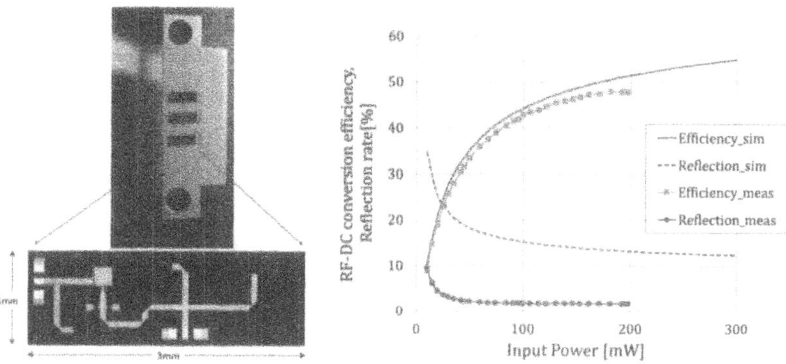

a) b)

Figure 3.83. *a) Developed MMIC rectenna at 24 GHz and b) RF–DC conversion efficiency [HAT 13]*

A 35 GHz rectenna was developed at Texas A&M University (Figure 3.84) [YOO 92, MCS 96]. The first 35 GHz rectenna was developed with 39% power conversion efficiency at approximately 100 mW and 400 Ω[17]. The researchers then modified the 35 GHz rectenna by replacing the dipole with a rectangular patch antenna and the modified rectenna showed an efficiency of 60 % at an input power of 25 mW based on a free-space measurement

[MCS 96]. Texas A&M University also simultaneously developed a 10 GHz rectenna with 60% efficiency.

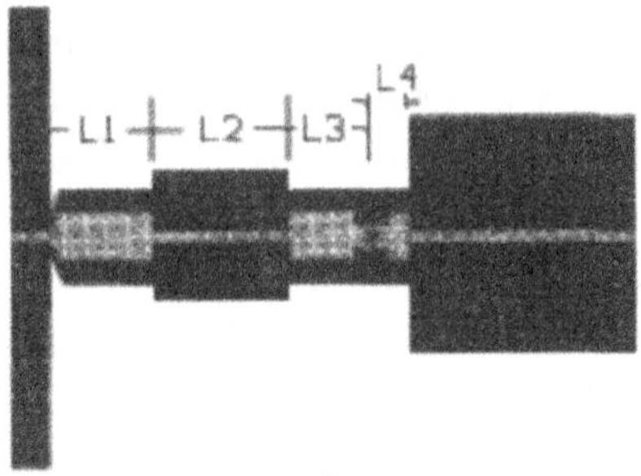

a)

b)

Figure 3.84. *a) Design of a 35 GHz rectenna with a dipole antenna and b) image of the developed rectenna [YOO 92]*

A 60 GHz rectenna was developed with MMIC technology at Eindhoven University of Technology, the Netherlands [GAO 13]. They indicated that several low-impedance paths caused by parasitic capacitances lead to a trade-off between isolation and insertion-loss at 60 GHz. Therefore, the inductor-peaked rectifier structure was proposed and realized (Figure 3.85). An inductor-peaked diode-connected transistor, self-threshold voltage modulation and a low-pass output filter are used to increase the sensitivity and the efficiency of the rectifier. The series input inductor can take advantage of the voltage boost effect. The designed rectifier reaches 7% efficiency at 62 GHz with –14 dBm input.

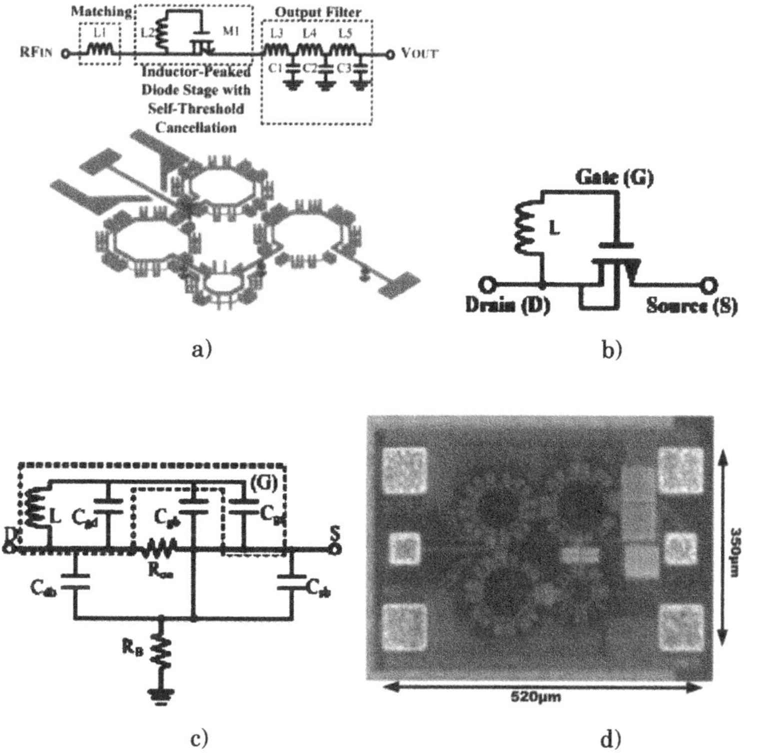

Figure 3.85. *Rectenna of 60 GHz. a) Schematic of inductor-peaked rectifier and layout model for electromagnetic field simulation; b) inductor-peaked diode-connected transistor; c) circuit model in the on-state; d) die micrograph of 62 GHz inductor-peaked rectifier[GUH 13]*

Various other RF and microwave harvester modules have been reported in the literature: amplitude modulation (AM broadcasting radio [WAN 13], broadcasting TV [SAM 09, KAW 13], Global System for mobile communications (GSM) [POW 13], global positioning system (GPS) [ZHU 11], X-band [EPP 00] and K-band [TAK 13].

Dual-band rectennas are useful for various applications. A 2.45 and 5.8 GHz dual-band rectenna, in which a single-shunt rectifier was used, was developed at Texas A&M University (Figure 3.86) [SUH 02].

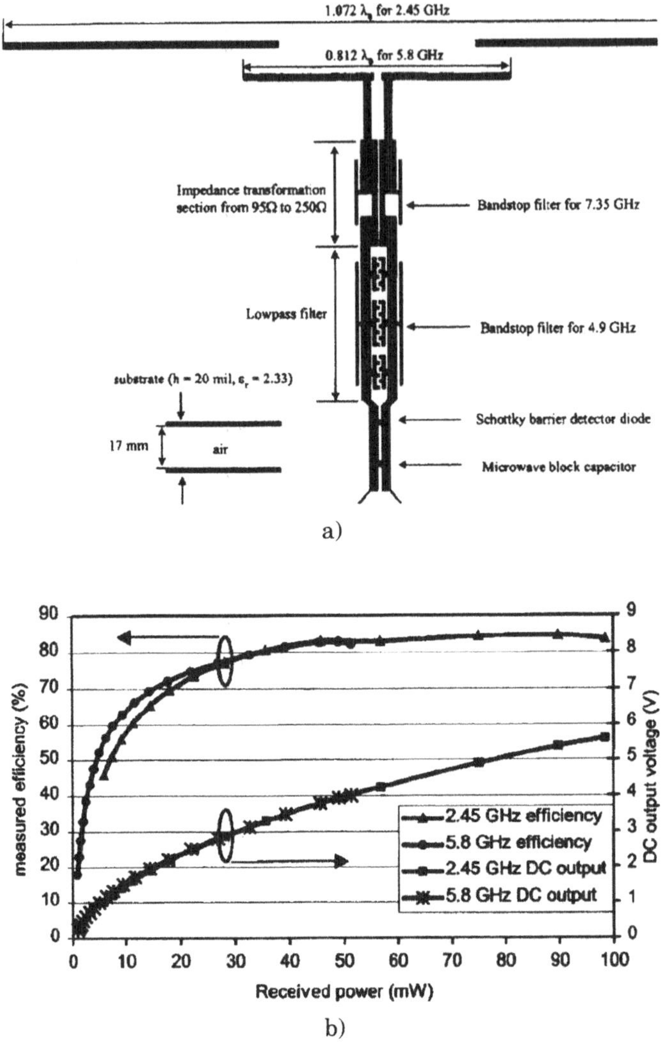

Figure 3.86. *a) Dual-band rectenna developed at Texas A&M University and b)RF–DC conversion efficiency and reflection ratio of dual-band rectenna [SUH 02]*

A triple-band rectenna (900 MHz, 1.9 GHz and 2.4 GHz) was also developed at the University of California, Davis (Figure 3.87) [PHA 13]. The antenna was designed using a combination of three different design techniques including a composite right/left-hand transmission line. For the rectifier, which was of the charge-pump type, they tuned each matching frequency independently.

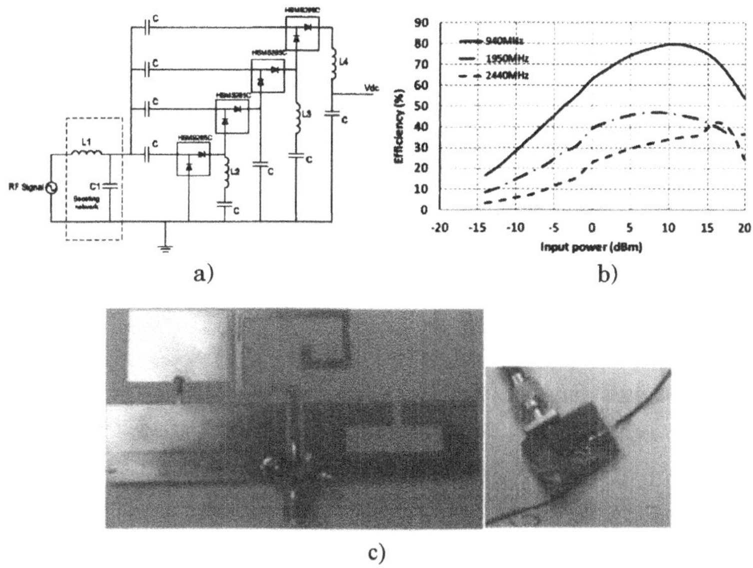

Figure 3.87. *Triple-band rectenna developed at UC Davis. a) Schematic diagram of proposed rectifier, b) RF–DC conversion efficiency and c) image of triple-band antenna and rectifier [PLA 13]*

3.5.4. *Various rectennas III – weak power and energy harvester*

For energy harvesting, a high-efficiency rectenna capable of harvesting weak power signals is required. It is also important to realize an MPT to sensor network or weak power MPT applications suitable for an initial commercial phase. As mentioned in section 3.2, a rectenna with a diode usually cannot realize high efficiency at weak power. The V_J effect cannot increase the efficiency. So we should consider how we can add a higher voltage to V_J on a diode without any extra power source.

There are some high-efficiency rectennas operating at weak power, and they use various circuits as follows:

1) Charge-pump rectifier [POW 13, PHA 13, MOR 06, YAN 13].

2) Rectifier with a resonator [KIT 06, YAM 13].

3) Rectifier using reflected microwaves [UED 04].

4) Rectifier with the design of output filter [SHI 04, SHI 13b].

5) Analytical model with consideration of higher harmonics [AKK 05].

The charge-pump rectifier (Figure 3.88) is currently in common use for energy harvesting or weak power MPT because the required diode voltage can be created. But its RF–DC conversion efficiency is generally low because many diodes and many capacitances are needed to increase the voltage.

To increase voltage on a diode, some groups have used a resonator. Tohoku University's group used a short-stub resonator (Figure 3.89(a)) [KIT 06]. They achieved an RF–DC conversion efficiency of ~40% at 900 MHz and 10 μW. Toyama University's group adopted an L–C resonator and a dielectric resonator (Figures 3.89(b) and (c)) [YAM 13].

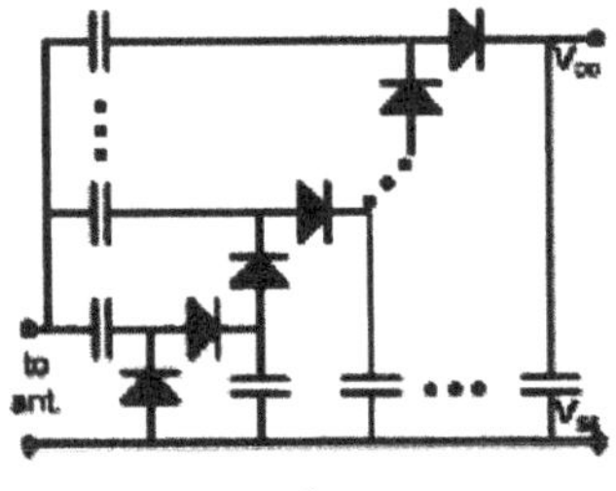

a)

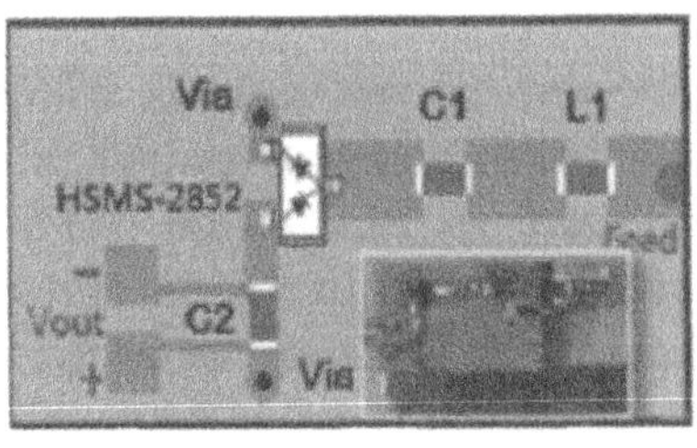

b)

Figure 3.88. *Charge-pump rectifier. a) Schematic and b) implementation [YAN 13]*

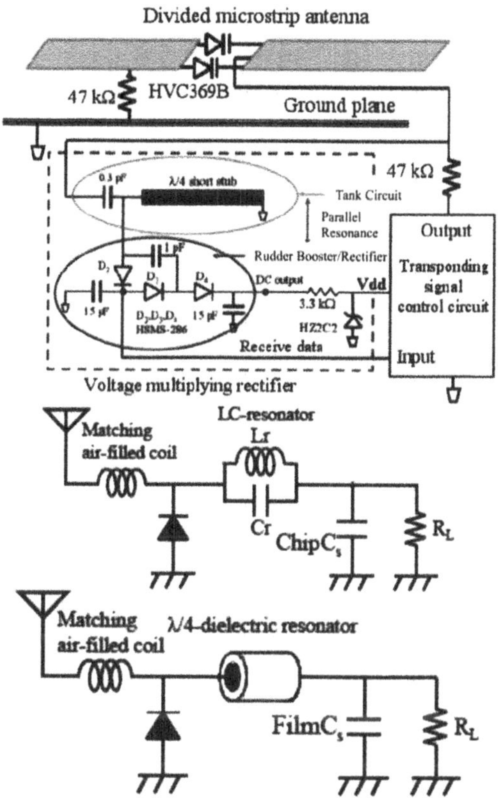

Figure 3.89. *Rectifier with a resonator. a) With a short-stub [KIT 06], b) with an L-C resonator [YAM 13] and c) with a dielectric resonator [YAM 13]*

Utilization of the reflected microwave and the standing wave is a feasible means of increasing the voltage on a diode. Okayama University's group tried to develop a system based on this concept [UED 04]. They used the reflected microwave from an output filter. Kyoto University and Mitsubishi Electric Corp. also focused on the output filter to increase RF–DC conversion efficiency. For the output filter, a distributed line length of $\lambda_g/4$ must be used to realize a full-wave rectifier with a single shunt [GUT 79]. But we also searched for the optimum highest length of the distributed line on the output filter to achieve the highest RF–DC conversion efficiency at 1 mW and 5.8 GHz. We found that <λ_g/10 distributed line is optimal (Figure 3.90(a)) and we experimentally achieved more than 50% efficiency at 1 mW and 5.8 GHz. For a 2.45 GHz rectenna, we additionally considered the number of diodes as shown in Figure 3.90(b). Finally, we achieved 55.3% efficiency at 0.1 mW and 8.2 kΩ at 2.45 GHz in a circuit simulation.

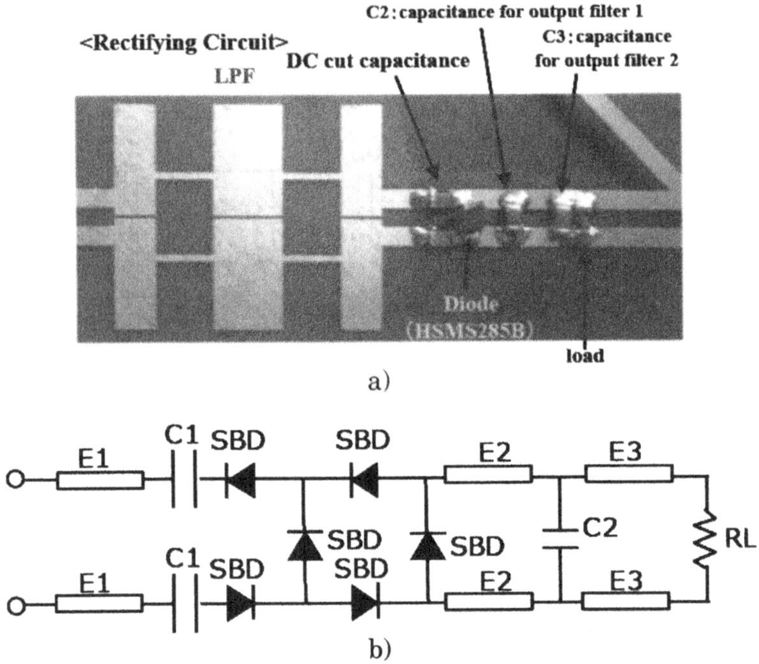

Figure 3.90. *Rectifier with the design of an output filter. a) With < λ_g / 10 distributed line at 5.8 GHz [SHI 04, SHI 13] and b) with optimization of number of diodes and length of distributed lines at 2.45 GHz [SHI 13]*

A zero-bias diode is a device which could potentially be used to increase the RF–DC conversion efficiency for weak microwave signals. Some trials have been conducted using a zero-bias diode but the RF–DC conversion efficiency is still low [TAK 13, ZBI 06].

3.5.5. *Rectenna array*

A rectenna will normally be used as an array for high-power MPT because one rectenna element only rectifies a few W. Although there are many existing studies of rectenna elements as discussed in previous sections, only a few rectenna arrays have been developed and used for experiments (such as the array shown in Figure 3.91). The largest rectenna array in the world was used for a ground-to-ground experiment in Goldstone, USA, by JPL in 1975 as shown in section 1.2. The size was 3.4 m × 7.2 m = 24.5 m^2. A rectenna array that had 2,304 elements and was 3.54 m × 3.2 m in size was developed for a ground-to-ground experiment conducted by Kyoto University, Kobe University, and Kansai Electric Corporation in 1994. Kyoto University has several types of rectenna arrays designed for 2.45 and 5.8 GHz operation. The area of these arrays is approximately 1 mϕ. Another rectenna array with a size of 2.7 m × 3.4 m was developed for a fuel-free airship MPT experiment conducted by CRL and Kobe University in Japan in 1995. A large rectenna was developed for a Reunion Island project, as previously shown in Figure 1.9.

For usual phased array antenna applications for radar systems, etc. mutual coupling and phase distribution are the problems which need to be solved. For the rectenna array, the problem is different from that of the array antenna because it is connected in the DC domain that includes no phase. Figure 3.92 shows the beam pattern of an antenna element that is assumed to be a cosine pattern for an array with normal antennas and an array with rectennas, respectively. Maximum power is normalized for each array. The beam pattern becomes narrower when antenna elements are connected in an array. But the beam of the rectenna array does not become narrower and is the same as that of one antenna element. This is due to the DC connection in the rectenna array where phase coupling is not an issue.

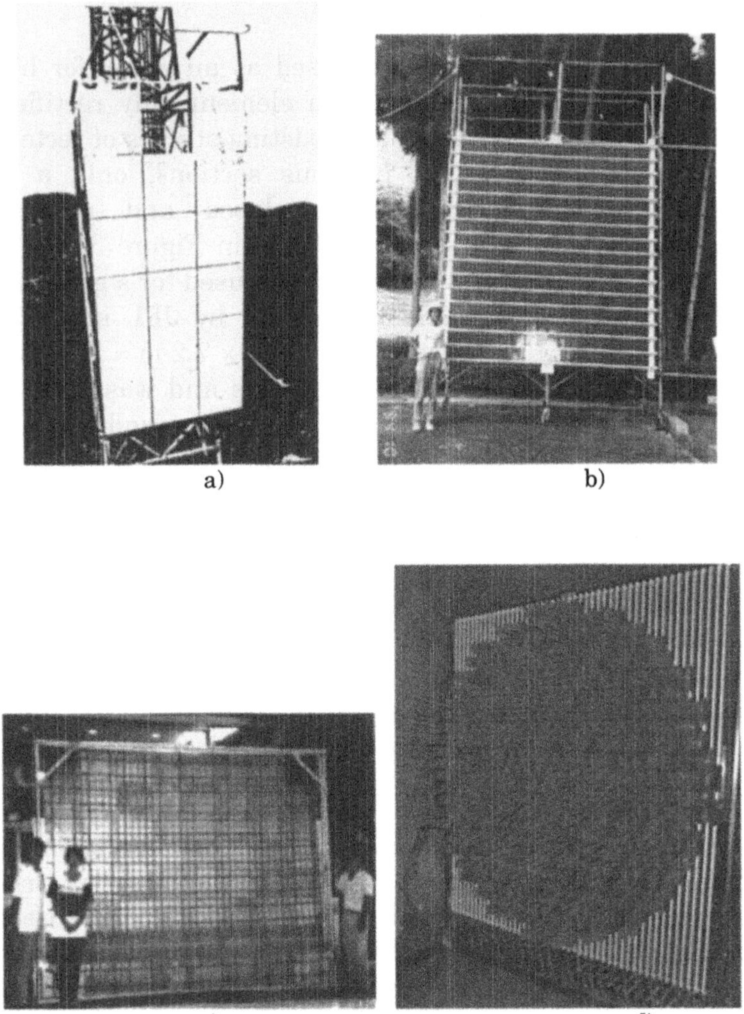

a) b) c) d)

Figure 3.91. *Large rectenna arrays used for a) G-to-G experiment in Goldstone in 1975, b) G-to-G experiment in Japan in 1994–1995, c) fuel-free airship experiment in 1995, d) experimental equipment in Kyoto University*

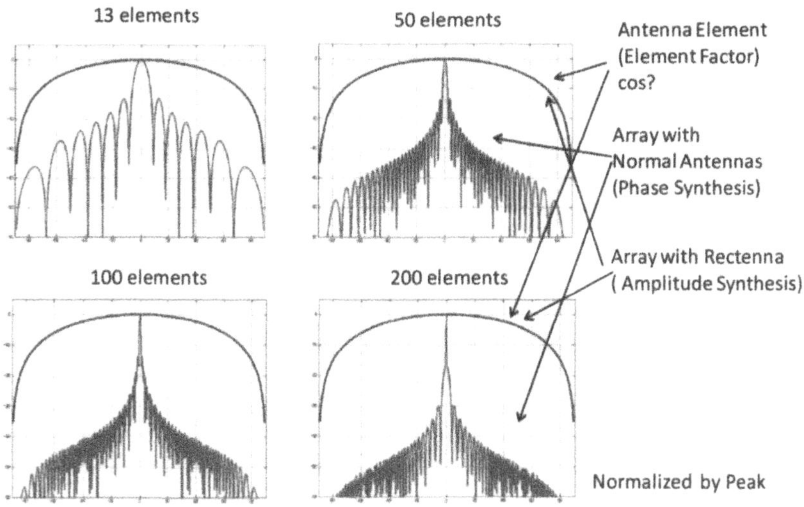

Figure 3.92. *Cosine beam pattern of antenna element, array with normal antennas and array with rectenna*

A rectenna has the characteristics of RF–DC conversion efficiency shown in Figure 3.69. It depends on an optimum connected load to achieve maximum RF–DC conversion efficiency at a given microwave frequency. The characteristics of the RF–DC conversion efficiency can be explained by the maximum power transfer theorem based on the equivalent circuit of a rectifying circuit that consists of a DC source and an internal resistance (Figure 3.93). When $R_L = R_S$, power supply is at the maximum and corresponds to the maximum RF–DC conversion efficiency. The internal resistance is not a real resistance. V-I curve of a diode as shown in Figure 3.70 creases the characteristics of the RF-DC conversion efficiency of the rectenna which can be explained with the 'image' internal resistance. In many rectennas, R_S is approximately several hundred Ohms, depending on the V-I characteristics of the Si Schottky barrier diode or other diode types. This means that the rectifying circuit in a rectenna is considered as a theoretical circuit with a DC power source feeding an internal resistance of several hundred Ohms. A load of several hundred Ohms is too small for a normal DC power supply, so we always have to consider the optimum load to apply the rectenna.

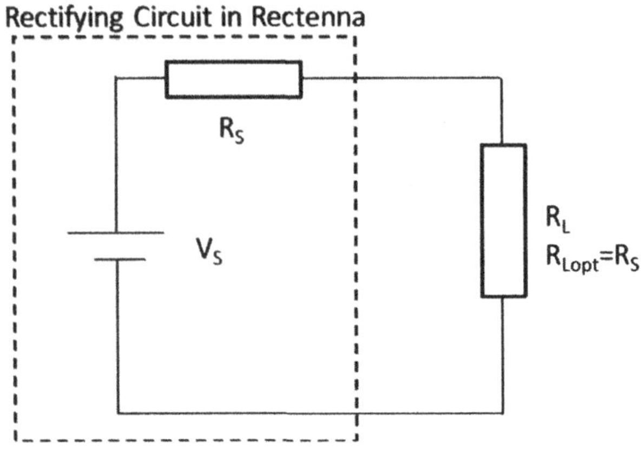

Figure 3.93. *Equivalent circuit of rectifying circuit*

When two rectennas are connected in series or in parallel, the optimum load requirement changes. Figure 3.94 shows experimental results of a rectenna array conducted in 1994–1995 by Kyoto University [SHI 98b]. In this experiment, 256 rectenna panels composed of nine rectennas connected in parallel were used, and the connections of the rectenna panels were changed from 128 in series and 2 in parallel to 8 in series and 32 in parallel. Figure 3.94 shows the load dependence of the reconfigured array, where the horizontal axis indicates the normalized resistance. The normalizing resistance R_N was predicted by theoretical calculations based on the equivalent circuits shown in Figure 3.95. If two rectennas are connected in series, the optimum resistance of the array becomes $R_{S1} + R_{S2}$. If two rectennas are connected in parallel, the optimum resistance becomes $(1/R_{S1} + 1/R_{S2})^{-1}$. These simple theoretical results are supported by the experimental results shown in Figure 3.94. This means that the optimum load can be predicted on the basis of how rectennas are connected in an array.

When the amplitude of the input microwave power supplied to two connected rectennas is different, for example $V_{S1} >> V_{S2}$, they will not operate at their optimum points and their combined DC output would be less than if they were operated independently. This can be explained by the characteristics of RF–DC conversion efficiency. It has been experimentally and theoretically proven that the total DC output decreases with a series connection to a greater extent than with a parallel connection [SHI 98b]. It has been further confirmed by simulation and experiments that current equalization in a series connection is worse than voltage equalization in a parallel connection [MIU 01]. Thus, parallel connections are the optimum connections for rectenna arrays receiving microwave signals with significant differences in power levels.

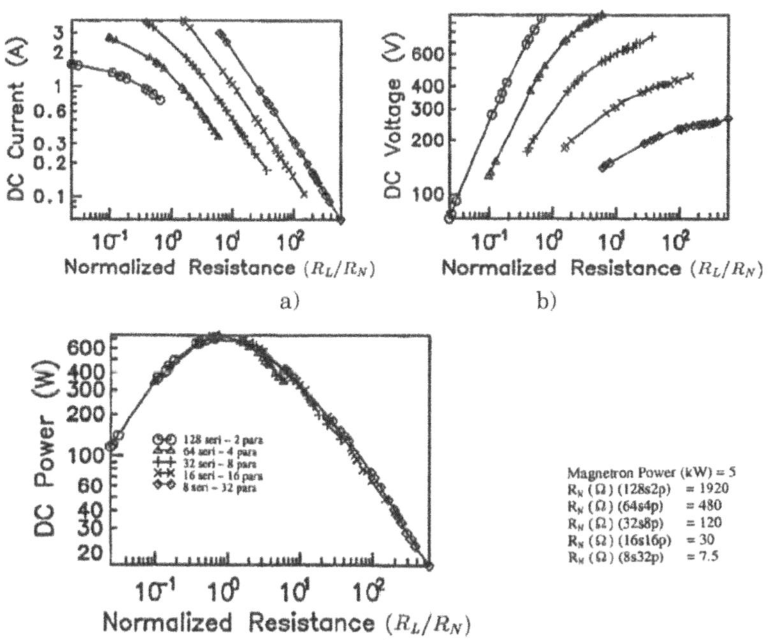

Figure 3.94. *Connected load dependence of a rectenna array; a) output DC current, b) output DC voltage and c) output DC power [SHI 98b]*

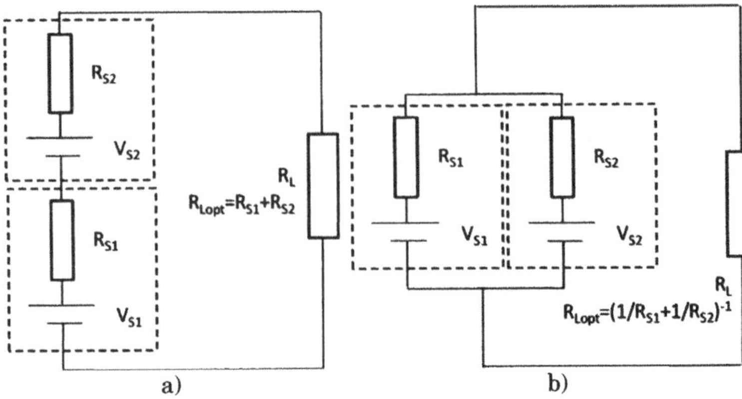

Figure 3.95. *Equivalent circuits of rectifying circuit: a) connected in series and b) connected in parallel*

3.5.6. *Rectifier using vacuum tube*

We can not only use a rectenna as a rectifier, but in certain situations we can also use a vacuum tube. If we would like to use a parabolic antenna as an MPT receiver, we use a cyclotron wave converter (CWC) instead of a rectenna. The CWC is a microwave vacuum tube used to rectify high-power microwaves directly into DC power. The most studied CWC consists of an electron gun, a microwave cavity with uniform transverse electric field in the interaction gap, a region with a symmetrically reversed (or decreasing to zero) static magnetic field and a collector with depressed potential, as shown in Figure 3.96. Microwave power from an external source is converted by this coupler into the energy for the electron beam rotation, and the latter is transformed into additional energy for the longitudinal motion of the electron beam by a reversed static magnetic field; then, the energy is extracted by decelerating the electric field of the collector and appears as a DC signal at the load resistance of this collector.

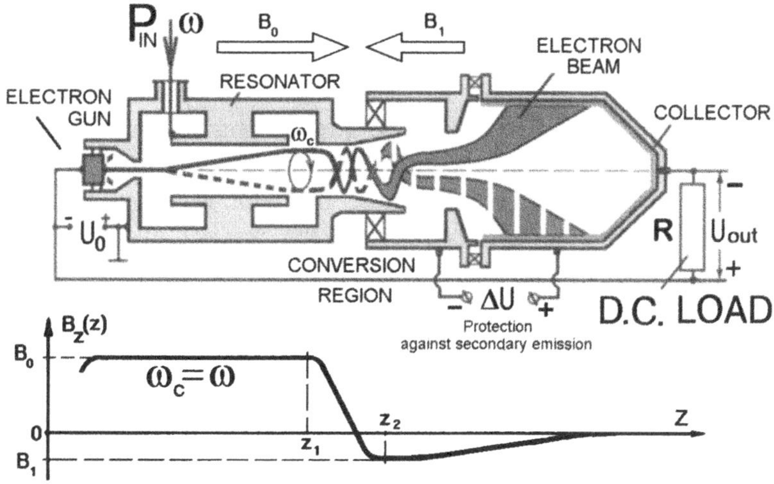

Figure 3.96. *Schematic of a cyclotron wave converter*

The first CWC experiment was carried out by Watson *et al.* [WAT 68, WAT 66, WAT 70]. The first CWC could only rectify 1–1.5 W input with 56% efficiency. At Moscow State University, a variant of the CWC was tested and its efficiency was 70–74% at 2.5–25 W. The TORIY Corporation and Moscow State University collaborated to create several high-power CWCs with efficiencies of 60–83% at 10–20 kW [VAN 82, VAN 91, VAN 95]. They demonstrated the CWC at the WPT '95 conference in Kobe, Japan. Vanke's group has continued to improve the CWC (Figure 3.62) [VAN 98, VAN 03]. A European group plans to apply a CWC for a ground-to-ground MPT experiment on Reunion Island [CEL 04], but this project has not yet been carried out.

Brown also proposed the use of a vacuum microwave rectifier in the 1960s [BRO 64]. He proposed a rectifier based on multipactor discharge. The term "multipactor" is derived from the words "multiple electron impact". A multipactor discharge consists of a thin electron cloud that is driven back and forth across a gap in response to an RF field applied across the gap. For the discharge to be self-sustaining, the second emission coefficient of the surface of the electrodes must be greater than unity, and the magnitude and frequency of the applied RF field must be adjusted so that the electron cloud traverses the gap during the half-cycle of the RF field. Under these conditions, successive impacts of the electron cloud produce a greater electron density, and the secondary electrons created with each impact are driven to the opposite electrode, producing additional secondary charge. Thus, the discharge current builds up. As the electron density increases, however, mutual repulsion causes some of the electrons to fall out of step with the field and thus limit the maximum electron density to a value controlled by the secondary emission coefficient [FOR 59].

Figure 3.97. *CWCs developed in Russia [VAN 03]*

Multipactor discharges can be used to provide both half-wave and full-wave rectification. A full-wave rectifier utilizes a re-entrant microwave cavity with the electric field concentrated at the center of the cavity where the secondary emitting electrodes are located (Figure 3.98) [BRO 64]. A single-cavity full-wave rectifier and a dual-cavity two-phase full-wave rectifier are shown in Figures 3.99(a) and (b), respectively. If the DC load voltage has been set equal to the electron impact voltage, the RF–DC conversion efficiency of the single-cavity full-wave rectifier is theoretically 59%, and that of the dual-cavity two-phase rectifier theoretically approaches 92% or approximately $\pi/2$ times that of the single-cavity full-wave rectifier.

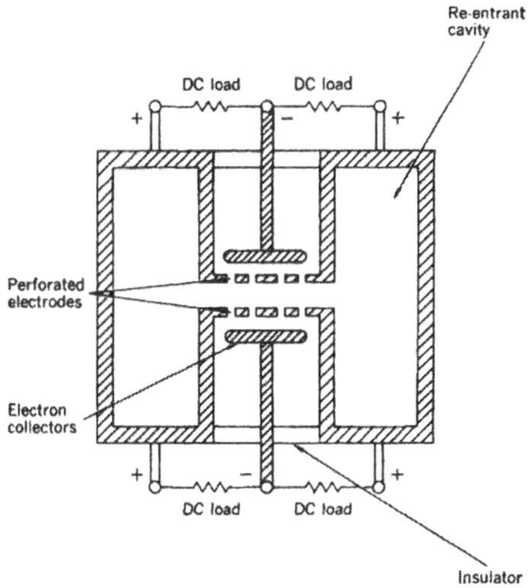

Figure 3.98. *Full-wave multipactor rectifier cavity [BRO 64]*

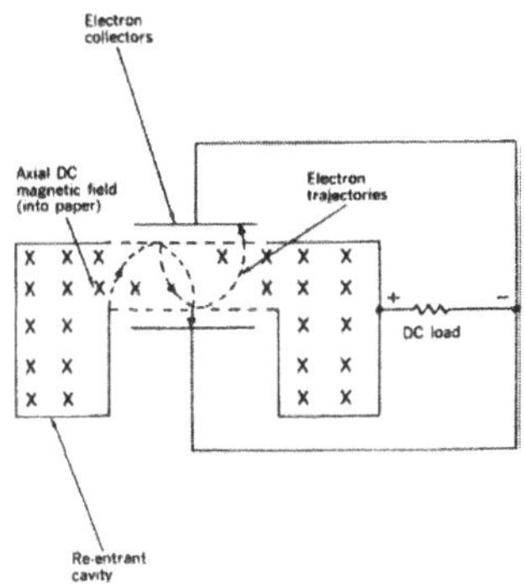

a)

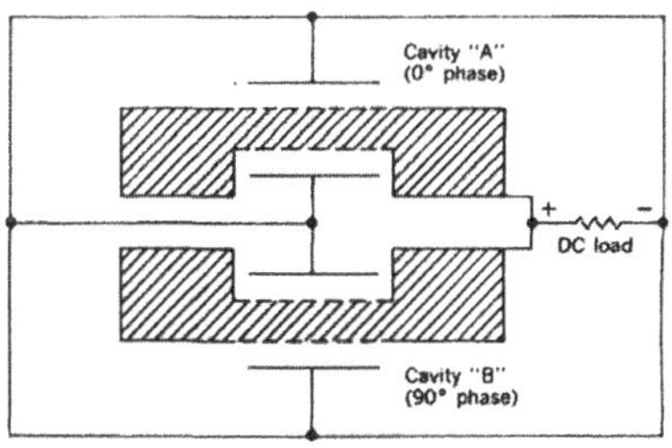

b)

Figure 3.99. *a) Single-cavity full-wave rectifier.*
b) Dual-cavity two-phase rectifier [BRO 64]

Chapter 4

Applications of WPT

4.1. Introduction

The following are descriptions of some of the main features of suitable applications of wireless power transmission (WPT) via radiowaves, especially microwave power transmission (MPT).

1) MPT using diffused radiowaves: MPT for multiusers using diffused radiowaves or energy harvesting from broadcasting waves (long-distance power transfer) (sections 4.2–4.4):

– similar wireless systems as used for information systems but with greater considerations for efficiency;

– coexistence of diffused wireless power and wireless communication;

– ubiquitous (everywhere and anytime) power source;

– possibility of utilization of lower frequencies with higher conversion efficiency because the radiowave source is diffused and is not a beam system;

– research topics: smaller size, large aperture antennas and high-efficiency low-frequency rectifiers.

2) MPT in a closed system or shorter distance MPT systems (sections 4.5–4.8):

– high-efficiency transmission with small antenna sizes over short distances or through closed waveguides;

– minimum unexpected radiation;

– free positioning of the transmitters and the receivers because of no coupling between the transmitters and the receivers;

– possibility of utilizing lower frequencies with higher efficiency;

– research topic: higher power semiconductors.

3) MPT using a beamed radiowave: MPT to a moving target from a transmitter (long distance power transfer) (sections 4.8–4.10):

– impossible to accomplish with wired power transmission, severely limited or no battery storage (for low weight applications) and with coupling WPT technology;

– high theoretical efficiency (>90%) over long distances;

– research topics: higher efficiency phased arrays and utilization of higher frequencies to enable reduction of antenna size.

4) Solar Power Satellites (SPS) (very long distance power transfer from geostationary orbits) (section 4.11):

– ten times larger power from solar cells in space (absence of nighttime and weather effects, etc.), total efficiency of 50% (beam MPT efficiency, conversion, etc.);

– research topics: lighter weight, higher efficiency, size reduction and lower cost phased arrays.

MPT for multiusers using diffused radiowaves or energy harvesting from broadcasted radiowaves competes with other energy harvesting sources, such as light, heat, wind and vibration. MPT in a closed system or short-distance MPT systems compete with inductive coupling and resonant coupling WPT. The MPT technology should be chosen with suitability of purpose in mind.

4.2. Energy harvesting

In the near future, if the power requirements of mobile devices sufficiently decrease, the power to operate mobile devices may derive from wireless power that can be received through broadcast service radiowaves. Such technology would be considered as a type of "energy harvesting". The energy harvester for broadcast radiowaves would be a rectenna used for both MPT and WPT. The rectenna has another merit for commercial applications in that energy harvesting is strictly a passive system without any active radiowave transfer and therefore requires no governing radiowave regulation.

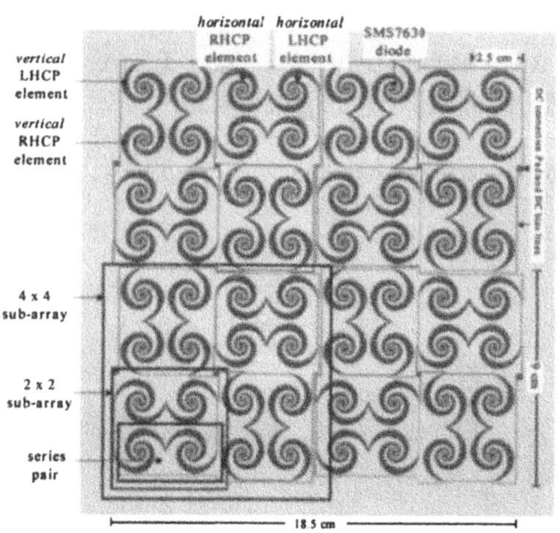

Figure 4.1. *64 Dual-circular-polarized spiral rectenna array active over a 2–18 GHz range with single-tone and multitone incident waves [HAG 04]*

In the early 21st Century, the University of Colorado began energy harvesting from an ambient radio frequency (RF) project which it called "recycling ambient microwave energy" [HAG 04]. Researchers developed a dual-circular-polarized spiral rectenna array operating in the frequency range of 2–18 GHz with single-tone and multitone incident waves whose power density was 10^{-5} to 10^{-1} mW/cm^2. The rectenna array developed is shown in Figure 4.1. They estimated that at least a 10 µJ or a 1 mW·µs load operation was required as a rectenna energy source. For example, it is for RF transmission of up to 10 kB data packets or simple sensory and signal processing functions. For the application, an average direct current (DC) power output through the rectenna with an RF–DC conversion circuit obtained a P_{DCmin} = 2 nW to P_{DCmax} = 450 µW when the incident power level was estimated as P_{RFmin} = 250 nW to P_{RFmax} = 2.5 mW and when the effective aperture of the rectenna element was approximately 25 cm^2 with an RF–DC conversion efficiency of the rectenna of 1% at P_{RFmin} to 20% at P_{RFmax} [HAG 04].

A US-based group, Intel Corporation, developed an energy harvesting system using channel 48 (674–680 MHz) and harvested 60 µW (0.7 V) with a Yagi–Uda antenna at a distance of approximately 4.1 km from the broadcasting station [SAM 09, SMI 10](Figure 4.2).

Figure 4.2. *Energy harvester designed by Intel Corporation [SAM 09]*

The Georgia Institute of Technology and the University of Tokyo also developed an energy harvesting system [VYA 12]. Table 4.1 indicates ambient wireless power measurements in Tokyo due to analog and digital TV broadcasts. Figure 4.3 shows ultra high frequency (UHF) and very high frequency (VHF) TV and radio broadcasts measured in Tokyo at a distance of 6.5 km from the Tokyo TV tower. In this radiowave application, the amount of wireless power required by the charge pump rectifier to charge the tank capacitor from 1.8 to 3.0 V was 38–70 W, respectively, at a frequency of 550 MHz, which is close to the wireless power captured by the antenna from UHF TV bands. The developed energy harvester is shown in Figure 4.4 and a photo of the field experiment taken in Tokyo is shown in Figure 4.5.

Frequency (MHz)	Application	Measured electric field intensity |E| (V/m)	Wireless power capture by dipole P_{IN} (μ-watts)
560–580	Digital TV	0.44–0.57	18.54–30.76
540–560	Digital TV	0.62–0.79	39.02–64.76
520–540	Digital TV	0.44–0.57	21.44–35.58
510–520	Digital TV	0.007–0.01	0.007–0.011
494	Analog TV	0.26–0.34	8.88–14.74
487	Analog TV	0.62–0.80	49.77–82.60
480	Analog TV	0.62–0.80	51.23–85.03

Table 4.1. *Ambient wireless power measurements in Tokyo due to analog and digital TV broadcasts [VYA 12]*

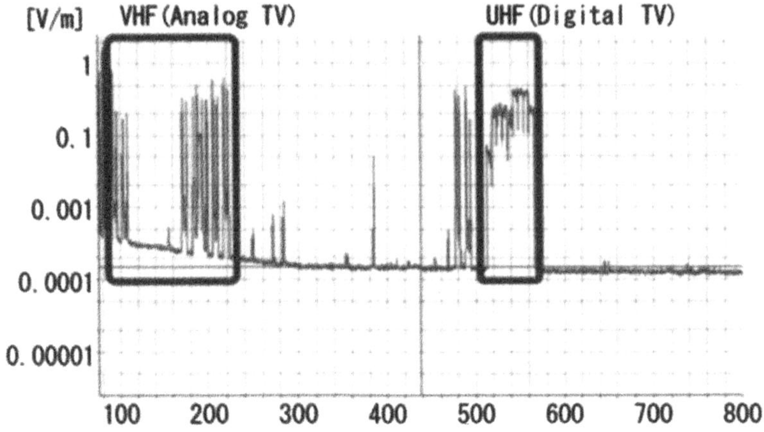

Figure 4.3. *UHF and VHF TV and radio broadcast power measured in Tokyo at a distance of 6.5 km from the Tokyo TV tower [VYA 12]*

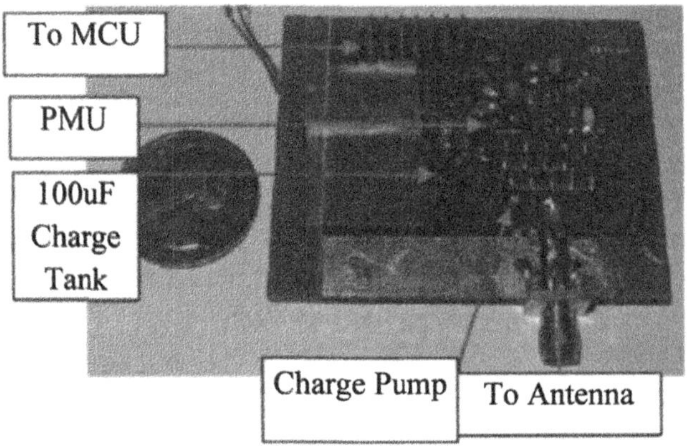

Figure 4.4. *Five-stage wireless energy harvesting RF charge pump circuit and PMU developed by GIT and University of Tokyo [VYA 12]*

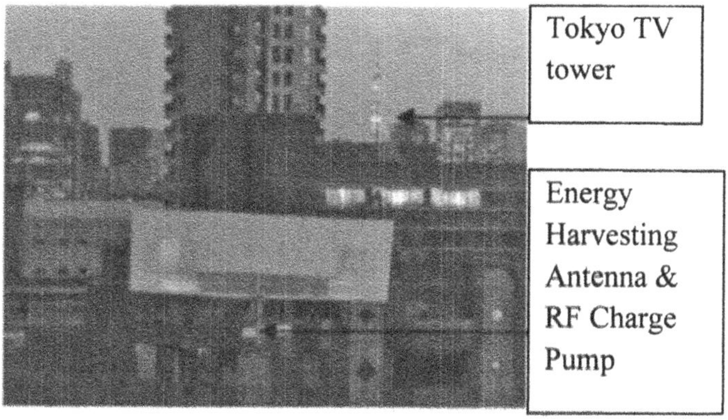

Figure 4.5. *Photo of the energy harvesting field experiment in downtown Tokyo [VYA 12]*

Another Japanese group carried out an energy harvesting field experiment in Nara, Japan [KIT 13]. The system was set up by the window of the researchers' office. The measured spectrum and power density are shown in Table 4.2 and Figure 4.6. In this application, the open-circuit voltage of the RF–DC conversion circuit was 450 mV and the received power in the 800 MHz band was measured at around –20 dBm using their developed twinloop antenna. A block diagram and photo of the developed system are shown in Figure 4.7. After a charging time of 19 h, the EDLC voltage charged up to 469 mV.

Source	Frequency	Received power	Power flux density
Mobile multimedia	207–222 MHz	–17.2 dBm	75.1 $\mu W/m^2$
DTV	470–560 MHz	–25.7 dBm	47.1 $\mu W/m^2$
Cellular BS	860–890 MHz	–26.8 dBm	133.7 $\mu W/m^2$
	2,110–2,170 MHz	–38.4 dBm	57.0 $\mu W/m^2$

Table 4.2. *Measured power density in Nara, Japan [KIT 13]*

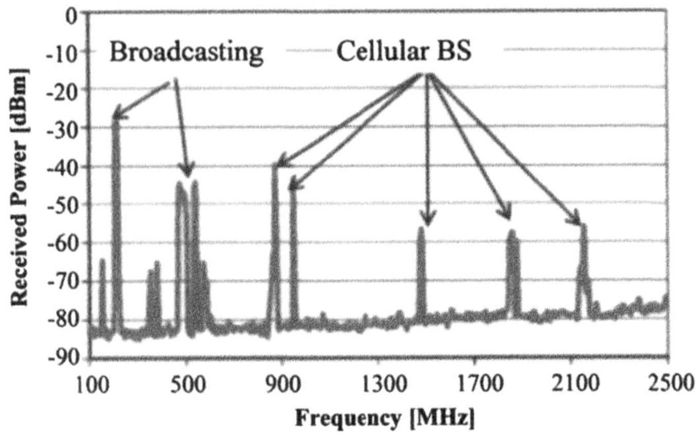

Figure 4.6. *Measured frequency spectrum in Nara, Japan [KIT 13]*

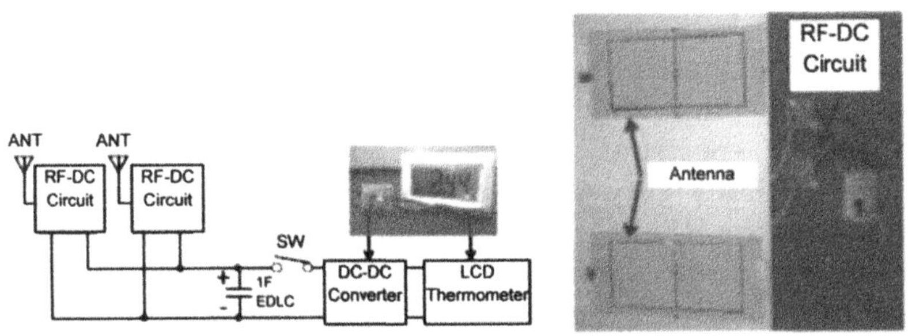

Figure 4.7. *Block diagram and photo of the energy harvesting system [KIT 13]*

An energy harvesting application has already been commercialized by Powercast Corporation in the United States [POW 13b]. The company released an energy harvester chip (Figure 4.8). The characteristics of the Powercast P2110 energy harvester are as follows:

1) designed for charging capacitors in battery-free devices or energy cells with high input impedance;

2) provides intermittent/pulsed power output;

3) harvesting range of 850–950 MHz;

4) works with standard 50 ohm antennas;

5) configurable, regulated output voltage up to 5.25 V;

6) power management and controlled input/output (I/O) for system optimization.

Figure 4.8. *Powercast energy harvester chip [POW 13b]*

The company has a patent for a rectifying circuit for wideband radiowaves (Figure 4.9) [ENE 06].

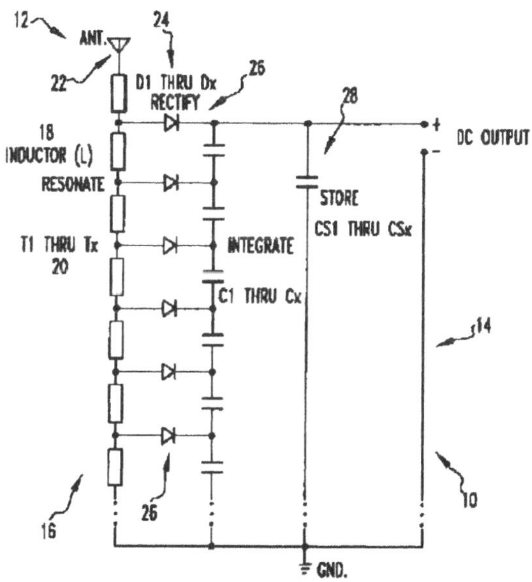

Figure 4.9. *Powercast patent for rectifying circuit for wideband radiowaves [ENE 06]*

4.3. Sensor network

If users do not require a large amount of power, microwave power can drive their devices without a battery. One potential low-power device is a wireless sensor. As such, wireless power can be actively transmitted to users that require electricity (Figure 4.10).

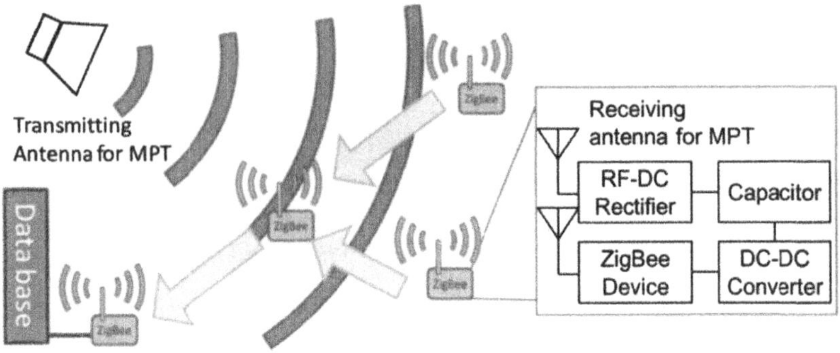

Figure 4.10. *Image of wireless powered sensors (ZigBee)*

The ZigBee/IEEE802.15.4 sensor represents a potential wireless powered sensor. The ZigBee network is configured as a coordinator, a router and an end device. The coordinator is just one device in the network and coordinates the network. The router has the function of relaying data from other routers and end devices as well as sensing information. The end device only sends data; otherwise, it sleeps. Therefore, power consumption of the end device is lower than the other devices. The router and end device send a total of nearly 2 ms of data every 1.14 s. The coordinator drops either device from its network if it does not receive data from them for 15 s. The out-of-network device must perform the necessary routines to rejoin the network. When the end device and the router rejoin the network and communicate with each other, they expend 9.46 and 57.4 mW, respectively, to do so. When they do not join the network and sleep, they expend 61.8 and 57.1 mW, respectively. The electricity consumption is one example that indicates we can drive the ZigBee sensor strictly with microwave power.

Hence, pulsed microwave power operating with the same frequency as ZigBee is suitable to increase microwave power by including suppression of interference with ZigBee's wireless communication (Figure 4.11). Experiments involving a microwave-driven ZigBee sensor and its coexistence with a pulsed microwave power and wireless communication system on the same frequency of 2.45 GHz were successfully carried out at Kyoto University, indicating that power and communication systems could operate without interference in the same environment and at the same frequency [ICH 12]. Figure 4.12 shows the experimental results of the minimum duty ratios of which the intermittent microwave power could drive the end device and error rates at the given duty ratios. The pulsed microwave with a pulse frequency of 10 Hz and a duty ratio of 0.4 could drive the ZigBee device without any interference with ZigBee communications.

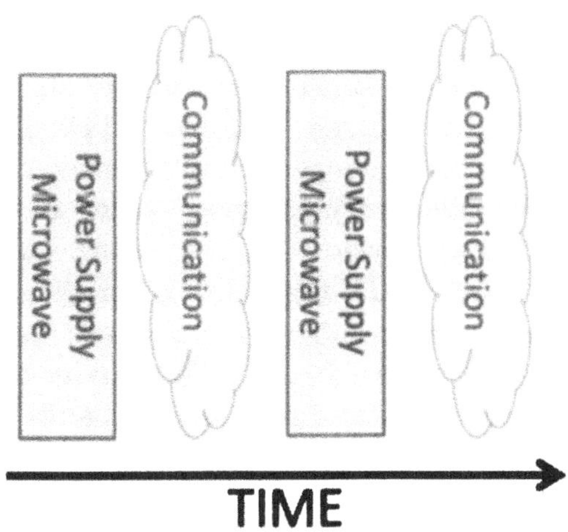

Figure 4.11. *Pulsed microwave power transmission whose frequency is the same as that of wireless communications*

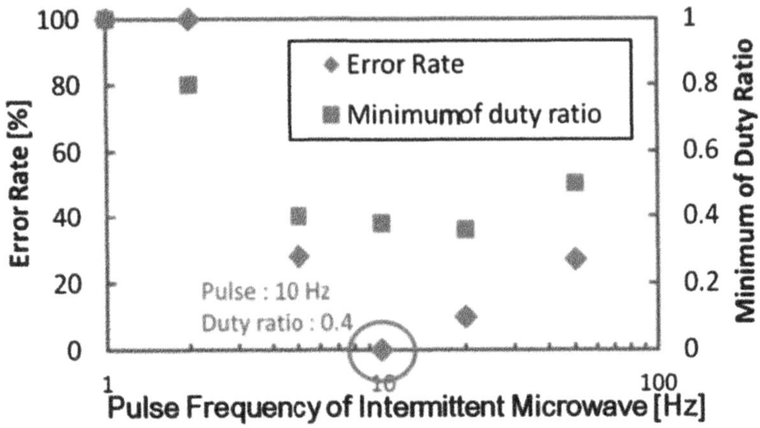

Figure 4.12. *Measurement results of error rates and minimum duty ratios with which the pulsed microwave could drive the end device [ICH 12]*

Another Japanese-proposed wireless powered sensor network is called wireless grid or microwave illumination – this concept is shown in Figure 4.13 [SAK 10, MAE 13]. The system provides wireless power and wireless information of active RFID in the 920 MHz band. The system uses four channels (1 W) without carrier sensing for passive RFID for wireless power. For the sensor network, 77 channels (1 mW) for active RFID are used (Figure 4.14). Instantaneous received power should be larger than the power consumed by the sensor. The average consumed power is controlled by changing the duty cycle (Figure 4.15). A multipower transmitter is used in the room and a carrier shift diversity is proposed in which multifrequencies are used to reduce standing waves in a closed room. The carrier shift diversity effectively reduced standing waves and created an approximately uniform power density in the room.

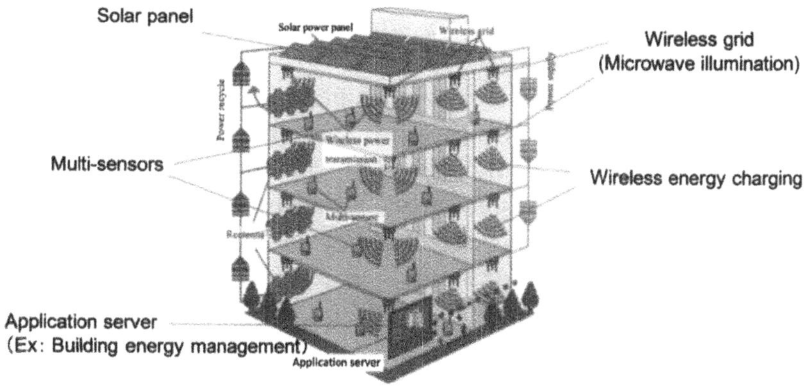

Figure 4.13. *Concept of the wireless grid (Microwave Illumination) [SAK 10]*

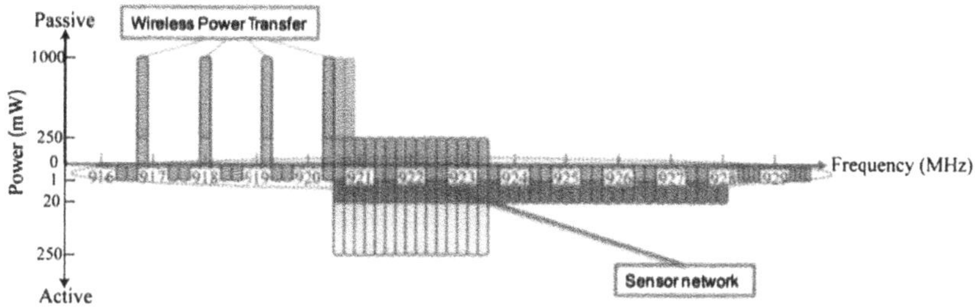

Figure 4.14. *Frequency spectrum for the wireless power and sensor network in the 920 MHz band [MAE 13]*

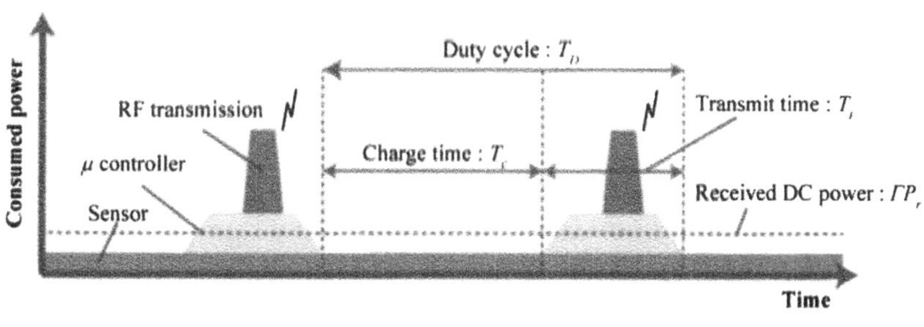

Figure 4.15. *Required power delivered to the activate sensor*

DENGYO Corporation in Japan provides a commercial wireless sensor using WPT in the 920 MHz band (Figure 4.16) [DEN 13]. The distance of application is less than 5 m. The RF–DC conversion efficiency is approximately 60%. The company proposes applications of the wireless sensor for sensing in high-temperature environments (85–120°C), sensing on a rotating or moving object and in severe environments, for example in outdoor or marine environments. DENGYO Corporation has developed a high-efficiency rectenna whose RF–DC conversion efficiency is approximately 86% at 2.45 GHz and 7 W [FUR 10].

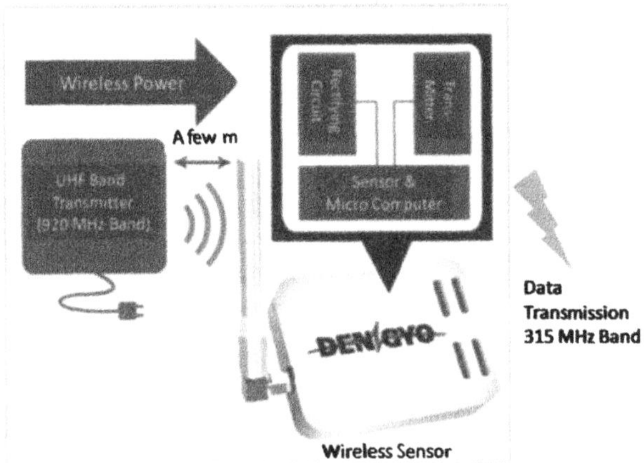

Figure 4.16. *Commercial wireless sensor using WPT in the 920 MHz band [DEN 13]*

4.4. Ubiquitous power source

Sensors are suitable devices to take advantage of WPT via radiowaves because they are very low power devices. However, various kinds of mobile devices exist that operate using battery storage and WPT should be applied for wireless power feeding or wireless charging of these various mobile devices. For this purpose, a idea named ubiquitous power source (UPS) was proposed near the end of the 1990s [SHI 04d, SHI 05] and it is based on the idea that microwaves are everywhere at all times (ubiquitous) and are always available for WPT.

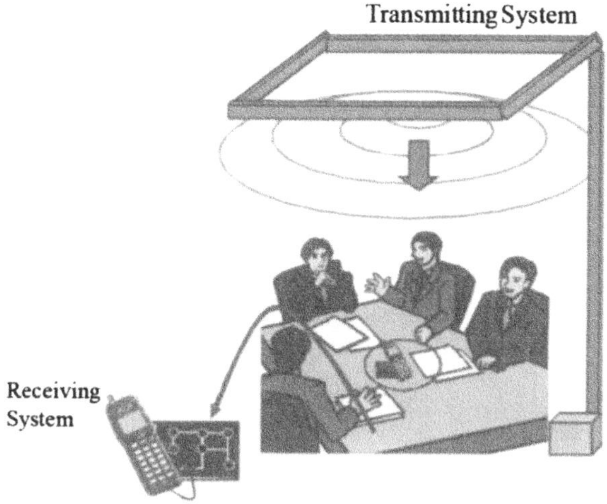

Figure 4.17. *The "ubiquitous power source" concept [SHI 04]*

The concept of UPS is shown in Figure 4.17. Microwave power at 2.45 GHz (industrial, scientific and medical (ISM) band frequencies in the 2.40–2.50 GHz range is transmitted from the edges of a ceiling to charge mobile phones. It is quite possible to create a uniform microwave power density in a UPS room with antennas installed at the ceiling edges. Slot antennas are selected as the transmitting antennas because of their reduccd cost. For the same reason, a frequency stabilized and phase controllable magnetron is used. However, because the UPS concept is based on a "wireless power source all the time and everywhere," the design of UPS is limited by safety issues associated with prolonged exposure to microwaves by human beings. The safety level set is under 1 mW/cm^2 for continuous exposure over the entire human body. For an experimental room of 5.8 m × 4.3 m, approximately 150 W of microwave power was transmitted from a magnetron to create a uniform microwave power density of 1 mW/cm^2 or less.

High-efficiency rectennas are also necessary for operation at a microwave power density of 1 mW/cm^2 or less. Under these conditions, an experiment was successful at charging mobile phones as shown in Figure 4.18. Furthermore, a mobile phone can still be used in the UPS room because of the difference between the 2.45 GHz of microwave power and the 1.9 GHz of communication system frequencies.

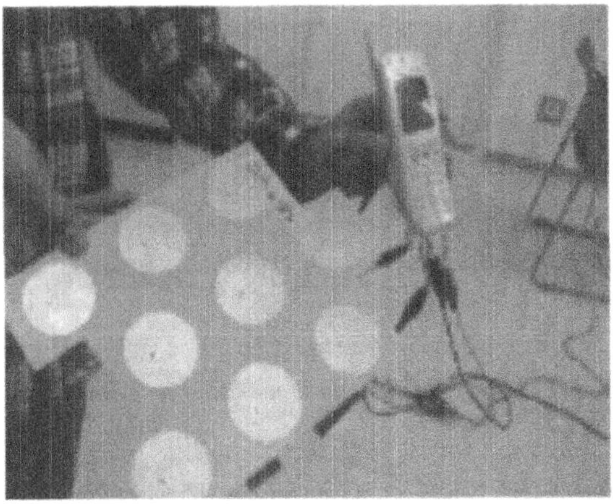

Figure 4.18. *Wireless mobile phone charging experiment using a UPS [SHI 04]*

As an extension of UPS systems, a phased array should be used to reduce the total transmitting power and to reduce unexpected radiation at positions where power is not needed (Figure 4.19). For directional UPS in a room the size of which is equivalent to the UPS experiment described above, only 22 W of microwave power (vs. 150 W for the conventional UPS system) is required if the power density around a device is to be 1 mW/cm^2. The phase array is still expensive for commercial WPT and UPS. The cost of the array depends on component costs, especially the phase shifter. Therefore, systems without phase shifters are proposed to reduce the total microwave power necessary for UPS while lowering the cost [HAS 11].

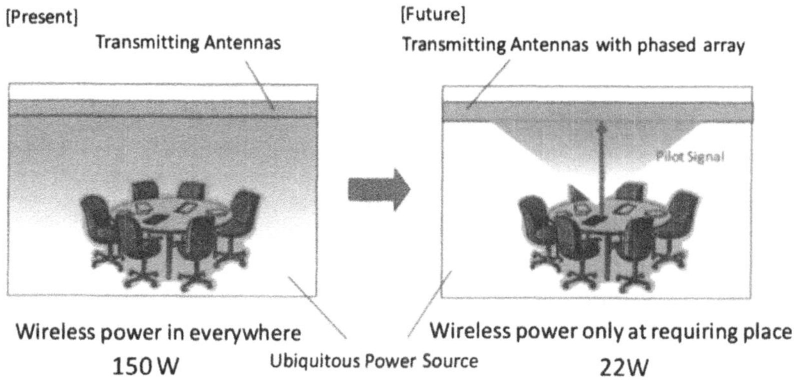

Figure 4.19. *Present UPS versus future UPS using a phased array*

An emergency UPS was proposed and an experiment was carried out in 2009 in Japan [MIT 10], as discussed in section 1.4. An image of the experimental system is shown in Figure 4.20. Worldwide, there are some research projects involving emergency base stations for mobile phones by balloon or by airship. However, even if a base station for mobile phones is established for emergencies, a mobile phone cannot be used without electricity. So an emergency UPS system is proposed for rapid and periodic recovery of electricity by wireless power. In the experiment of 2009, a mobile phone was charged using only wireless power from an airship (see Figure 1.16).

At the TechCrunch disrupt 2013 technology conference, Ossia Corporation, a US-based company, proposed a commercial wireless charger using MPT whose frequency is the same as that of WiFi [AOL 13]. It is called "Cota" and can wirelessly deliver 1 W of power at a distance of 30 ft. In the conference, they showed an iPhone 5 being remotely charged from a prototype WPT system. The company claimed that the commercialized version of Cota will be ready to ship in 2013–2014, and a consumer version will be ready to ship before 2015.

Figure 4.20. *Image of an emergency UPS system*

4.5. MPT in a pipe

MPT with diffused radiowaves must observe radiowave regulations in order to coexist with conventional wireless communications and other radiowave applications. Substantial wireless power cannot yet be transmitted under present radiowave regulations owing to questions of safety for humans and other living things exposed to radiowaves, where the 1 mW/cm^2 limit in the 2.45 and 5.8 GHz bands was described earlier as an example. The International Commission on Non-Ionizing Radiation Protection (ICNIRP) is the international policy group that makes safety guidelines for radiowave exposure [ICN 13].

To transmit substantial amounts of wireless power under present radiowave regulations, the concept of MPT in a pipe has been proposed and developed. The scheme has two merits. Microwaves and other radiowaves can propagate through a pipe, which serves as a waveguide, if the inside surface of the pipe consists of a conductor. In this situation, microwave power does not diffuse like it does in free space and only propagates through the waveguide. Theoretically, power loss for MPT over a distance D from a transmitter to a receiver in free space depends on the diffusion of the radiowaves in inverse proportion to the square of the distance D^2, which is determined by equation [2.5] or [2.7]. However, the propagation loss in the waveguide does not depend on the diffusion of the radiowaves but depends only on the product of the propagation distance D and a loss factor, which depends on the conductivity of the inside surface of the waveguide. This loss is much smaller than the loss incurred in free space propagation. Other merits include no interference with conventional wireless communication systems and no radiation-related safety problems for humans and other living organisms because the microwave does not propagate in free space. High-power microwaves can be easily transmitted only along the waveguide. The potential magnitude of microwave power that can be transmitted is not limited by interference and safety factors but by technical issues, especially the power limits of semiconductors and microwave circuits.

In a rectangular waveguide or a circular waveguide, a microwave propagates in a particular "mode" which is determined by Maxwell's equations using the boundary conditions (mainly the internal size and geometry) of the waveguide. There are a number of modes that decide the impedance of the waveguide and the resulting loss through the waveguide. Figure 4.21 illustrates a theoretical propagation mode in a circular waveguide. The loss coefficient α of a TE_{mn} mode wave and a TM_{mn} mode wave can be estimated theoretically. Figure 4.22 illustrates the theoretical loss of a circular waveguide. Based on the theory, we can apply this MPT as MPT in a pipe.

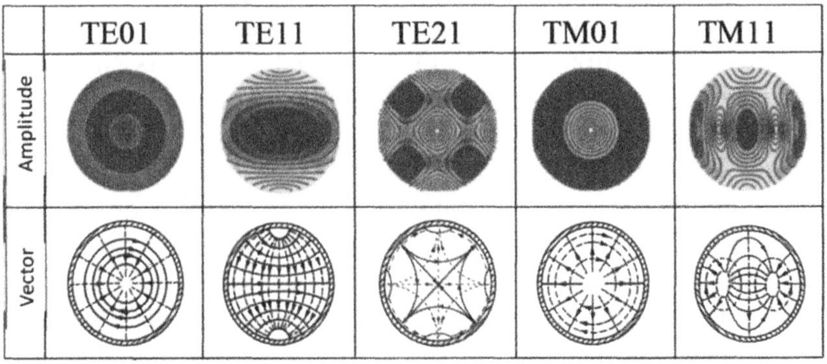

Figure 4.21. *Theoretical electromagnetic modes in a circular waveguide where the solid line denotes the electric field and the dotted line denotes the magnetic field*

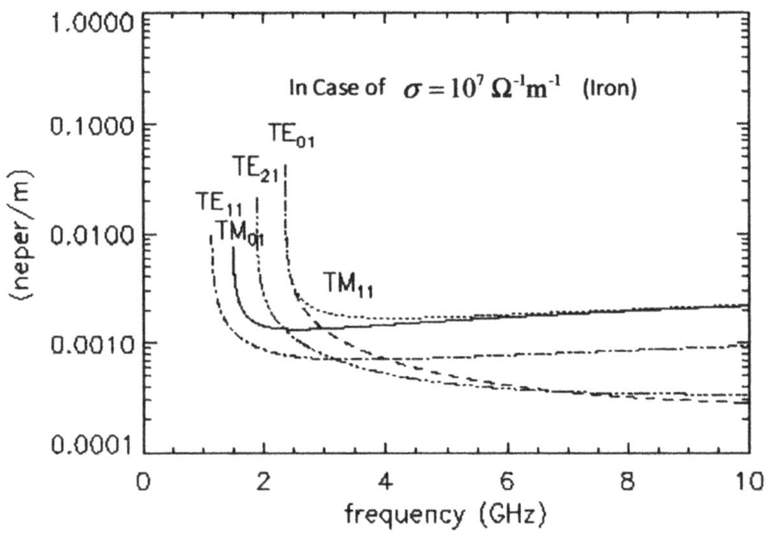

Figure 4.22. *Theoretical loss of a circular waveguide*

In the 1990s, in Japan, a microrobot moving in a pipe and powered by MPT in a pipe was developed. The concept is shown in Figure 4.23. Denso Corporation in Japan developed the MPT in a pipe system using 14 GHz microwaves, as shown in Figure 3.82 [SHI 97], propagating down a 15 mm diameter circular pipe, as a circular waveguide, in the TE_{11} mode (Figure 4.24) with a transmission loss estimated to be <1 dB/m. The rectenna which was composed of a monopole antenna and a rectifying circuit received the microwave power and fed the rectified DC power to the robot's inertial drive system composed of a piezoelectric bimorph cell to drive the microrobot. The microrobot received 50 mW of microwave power and was able to move at a rate of 1 mm/s in the pipe when 1 W of microwave power was transmitted through the pipe.

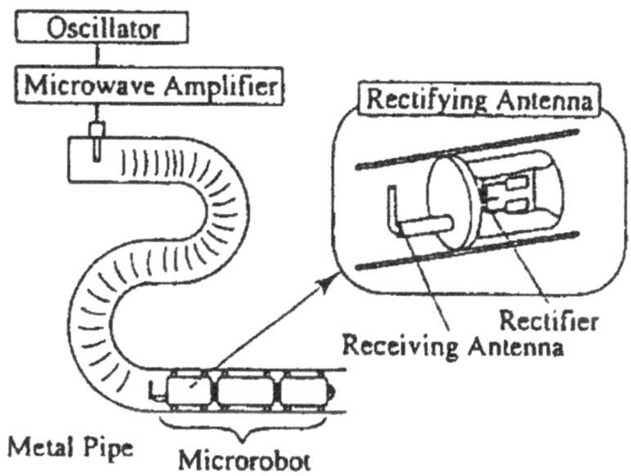

Figure 4.23. *The concept of a robot in a pipe powered by MPT [SHI 97]*

In the 1990s, Kyoto University and Osaka Gas Corporation also proposed an MPT system for powering an observation robot in a gas pipe [HIR 97, HIR 99]. The size of the gas pipe was approximately 155 mm, which was suitable for the propagation of microwaves operating at a frequency of 2.45 GHz. The problems associated with this particular MPT application were (1) an unknown propagation loss in rusty gas pipes and (2) the complex branching of the gas pipe network. The propagation loss was estimated by experiments and theory and was on the order of −0.1 to −1.0 dB/m, which depended on the propagation mode, and an estimated parameter for the electric conductivity $\sigma \cong 10^4 \Omega^{-1}\text{m}^{-1}$ of the inner surface of rusty gas pipes, and $\sigma \cong 10^7 \Omega^{-1}\text{m}^{-1}$ for an ideal iron pipe. It was concluded that the loss was sufficiently low to propagate microwave power in the gas pipe.

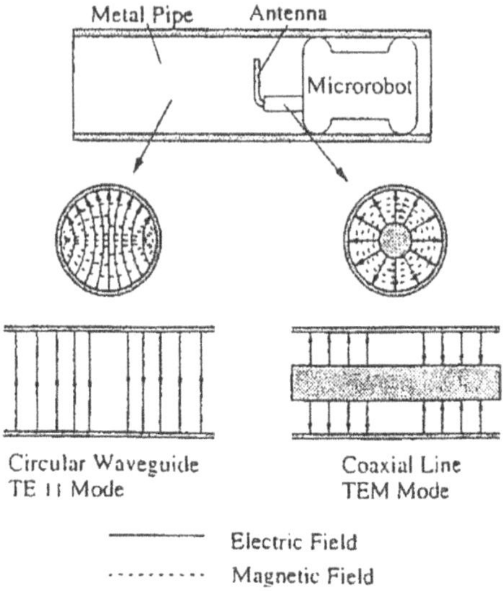

Figure 4.24. *Conventional method of power conversion from a circular waveguide to a coaxial cable [SHI 97]*

The complexity of the pipe branch network posed a bigger problem. Theoretically, radiowaves cannot propagate to all branches of the waveguide, and some branches would not be serviced with radiowaves. Experimental results supported the theory, which indicated that the proposed MPT in a pipe system had limited application.

4.6. Microwave buildings

Kashima Co., a Japanese building company, jointly proposed with Kyoto University a wireless building using microwave power technology [SHI 08]. The proposed power system is shown in Figure 4.25. This system wirelessly supplied electric power using a deck plate consisting of extra cover boards that acted as microwave transmission waveguides. A frequency of 2.45 GHz was selected on the basis of size limitations of the conventional deck plate, and a magnetron was used as the microwave transmitter to reduce cost. The flow of microwave power could be controlled by variable power dividers that supplied microwave power only to users requiring it and blocked flow to places where there were no users. Rectennas as DC converters and as DC power sources were placed under the floor. Adjusting the position of the rectenna was quite easy because microwaves were present practically everywhere under the floor. Total efficiency from electricity to DC via microwave transfer was assumed to be 50%. Although the day-by-day running costs of electricity for the microwave building system is approximately twice that of a conventionally wired home, the initial cost of the building is reduced because of reduced construction costs. Therefore, it was estimated that the overall lifecycle cost of the building can be reduced by using the microwave building system.

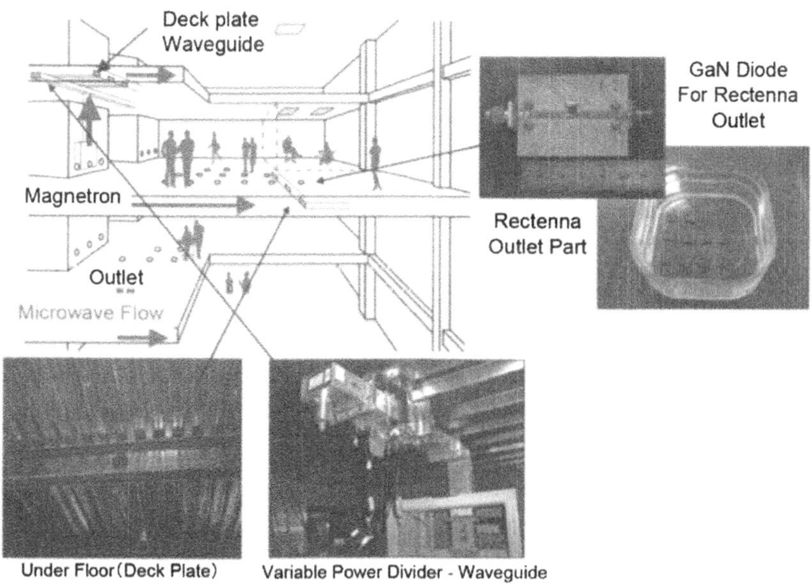

Figure 4.25. *Wireless building using microwave power transmission*

A result of an electromagnetic simulation of the conventional deck plate with the extra cover board using a finite element method is shown in Figure 4.26. The fundamental TE_{10} mode of the 2.45 GHz microwave is clearly shown. The cutoff frequency for the TE_{10} mode is 1.43 GHz and the cutoff frequency for the higher mode is 2.86 GHz. This indicates that 2.45 GHz microwave power can propagate through the conventional deck plate only with the extra cover board. The deck plate is plated with zinc and it is different with commonly used waveguides. The loss of the waveguide depends on the plating inside the waveguide. The experimental results describing the loss of the deck plate are shown in Table 4.3. The loss depends on the method of connection between the deck plate and the extra cover board. Use of the deck plate as the waveguide is sufficient to support the microwave building system.

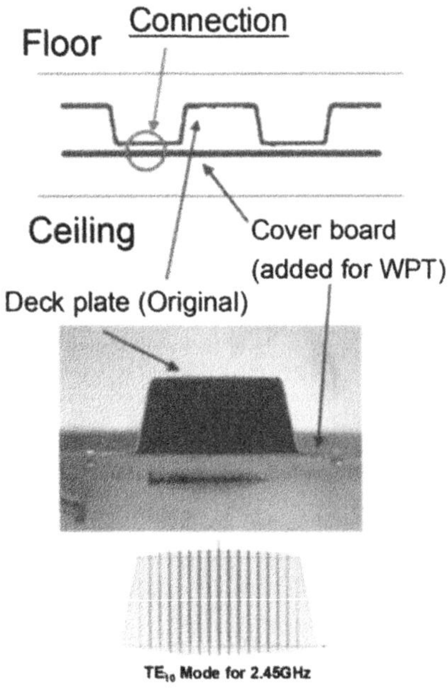

Figure 4.26. *Photo of the deck plate and simulation of the wave mode in the deck plate waveguide*

Connection	Attenuation constant [dB/m]
Spot welding	0.69
Bolted connection	0.37
Solder joint	0.02 (0.018 by theory)

Table 4.3. *Experimental results of power loss in the deck plate waveguide*

In this system, a high-efficiency rectenna with a high power input of at least 100 W is required as the power source. If the rectenna is developed with normal Si Schottky barrier diodes, over 250 diodes are required on a single rectenna. A high-power rectenna consisting of 256 diodes was developed at Kyoto University and its RF–DC conversion efficiency was approximately 55% at 100 W, 2.45 GHz microwave input. The size of the rectenna was 125 mm in length, 110 mm in width and 95 mm in height and it is too large to be used as a power source in the proposed application.

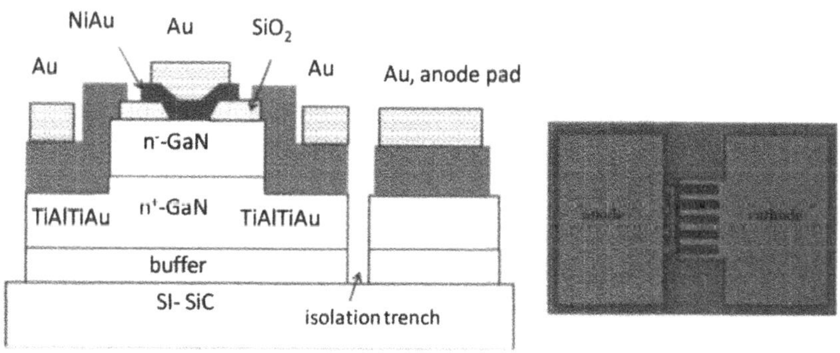

Figure 4.27. *A photograph of the developed GaN diode [TAK 09]*

GaN diodes are suitable for high-power rectification. As shown in Figure 3.3, GaN is applied for mainly HPA with HEMT and FET. There is still very little research on GaN diodes. In Kyoto University and Tokushima University in

Japan, a new GaN diode was developed for this application (Figure 4.27) [TAK 09]. The RF–DC conversion efficiency of the GaN diode rectenna developed is approximately 74.4% for 2.45 GHz, 5 W microwaves with a 130 Ω load (Figure 4.28). Such a rectenna would be suitable for the microwave building application.

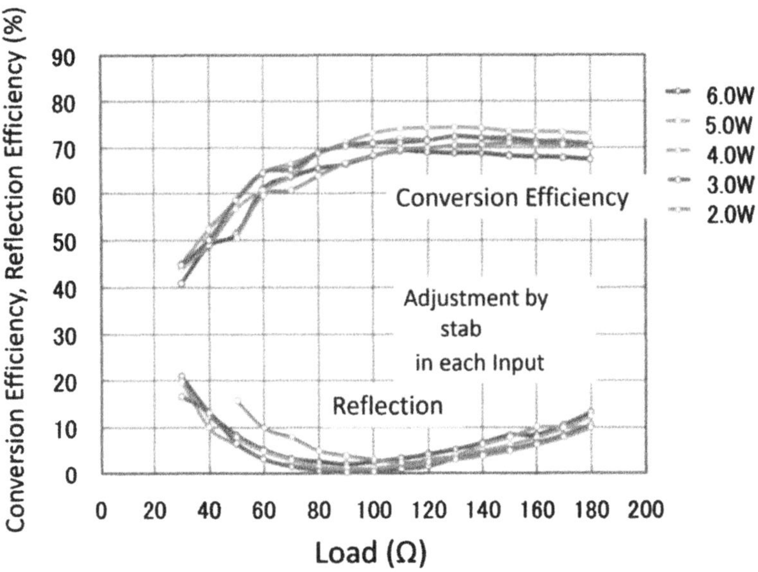

Figure 4.28. *The RF–DC conversion efficiency of the rectenna using GaN diodes. For a color version of this figure, see www.iste.co.uk / shinohara / radiowaves.zip*

In the initial phase, the wireless system was considered for office buildings where DC-driven computers and other OA instruments are mainly used. It is estimated that one DC converter requires <50 W, and >3 kW of microwave power is provided to a single room, indicating that the system provides sufficient power to run a number of typical electrical devices per room. The microwave traveling through the waveguide in this application serves as a UPS system.

4.7. 2D WPT

An MPT operating through closed waveguides is a good potential application under present radiowave regulations. Another MPT operating through closed waveguides was proposed and developed in Japan. It is called the "two-dimensional waveguide power transmission (2D WPT) system" [SHI 07a]. In 2D WPT systems, the microwave propagates along a waveguide sheet and is selectively received by special receiving devices on the sheet (Figure 4.29). The 2D WPT system also has an inevitable trade-off between the safety and power transmission capacity. Figures 4.30(a) and (b) illustrate an improved waveguide surface which has been demonstrated to enhance electromagnetic compatibility (EMC) performance [NOD 11, NOD 12b]. The microwave is received at a waveguide-ring resonator (WRR) coupler which extracts the power from the waveguide across a thick insulator (Figure 4.30(c)). Extraneous objects near or even touching the sheet are not exposed to the strong electromagnetic fields. The WRR coupler has a significantly high quality factor (high-Q) in the resonant state, which is essential to support the selective power transmission. The WRR is connected to a rectifying circuit which is driven in class F. The total efficiency, defined as the ratio of the DC output of the rectifying coupler to the microwave input into the sheet, was 40.4% at maximum, where the microwave input was 1 W in the 2.4 GHz band with a 6.4 × 3.6 cm^2 coupler on a 56 × 39 cm^2 sheet (nearly 100 times the area of the coupler) [NOD 12b].

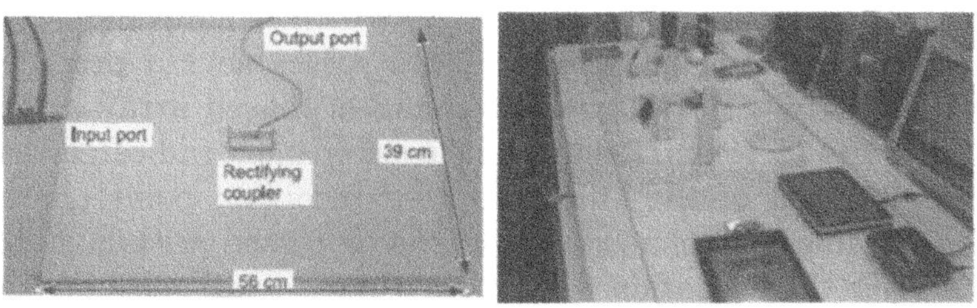

Figure 4.29. *Photographs of the 2D WPT system*

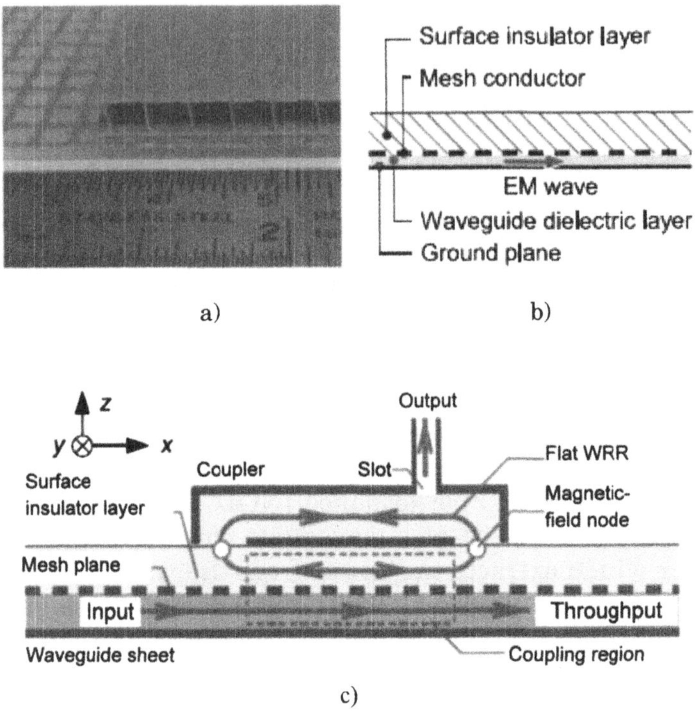

Figure 4.30. *Wireless power transfer sheet: a) upper view of the waveguide sheet around its edge, b) schematic drawing of its cross-section, c) receiver of wireless power [NOD 11, NOD 12b]*

Microwaves are used in the 2D WPT system and the microwave diffuses in accordance with Maxwell's equations even if the dimension decreases from three-dimensional (3D) to 2D. To suppress unexpected radiation, a phased array can be applied to the 2D WPT system, which is equivalent to a 3D WPT system [NOD 13]. Figure 4.31 shows a proposed experimental 2D WPT system using a phased array at the University of Tokyo. Results of the experiment indicated that the variation of the efficiency at a receiving point on the sheet using the phased array is suppressed within 2 dB, whereas the variation was more than 10 dB without the phased array system.

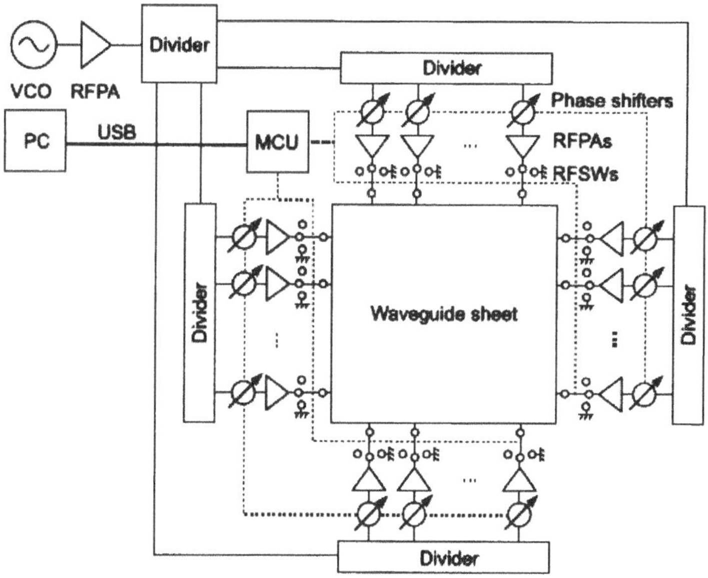

Figure 4.31. *Block diagram of the microwave power supply system using a phased array which has seven RF amplifiers along each edge [NOD 13]*

4.8. Wireless charging for electric vehicles

Highly efficient MPT can be applied not only through closed waveguides, but also over short distances with smaller antennas as shown in Figure 2.3. Photographs of the wireless charging of an electric vehicle (EV) using MPT are shown in Figure 4.32. It is convenient to apply MPT for wireless charging both for parked EVs and for EVs in motion because the transmitting and receiving antennas are not coupled. The impedance of the antennas is not changed by a change in the position of the EV nor is the efficiency of MPT changed. Problems of safety and interference by microwaves are diminished for both MPT in a closed waveguide and over short distances because there is almost no diffusion of unexpected microwaves. In short-distance power transfer systems, the magnitude of wireless power can also be increased to the kW range because the transmission does not interact with humans or other living organisms between the transmitting and receiving antennas.

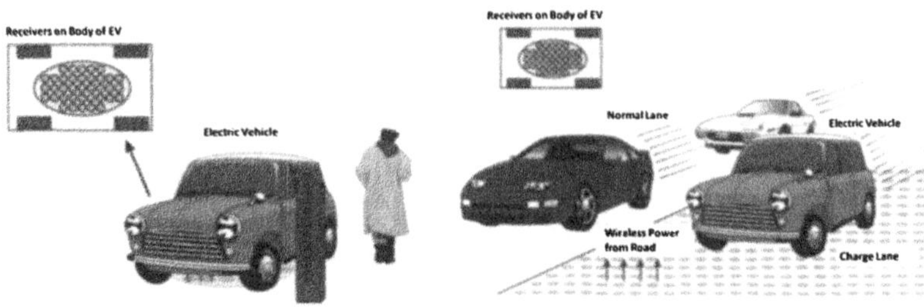

a) b)

Figure 4.32. *Images of wireless charging of EVs a) for parked EVs and b) for EVs in motion*

Kyoto University has proposed and implemented an MPT system for EVs (Figure 4.33) [SHI 04]. From 2003 to 2008, the university first collaborated with Nissan Motors to develop an MPT system between the road and the body of an EV using a microwave frequency of 2.45 GHz (Figure 4.34) [SHI 11a, SHI 11b]. Magnetrons and slot antennas were used to reduce the system cost. The distance between the transmitting and the receiving antennas was approximately 12.5 cm, a distance of 1 λ at the 2.45 GHz frequency. The battery on the electric vehicle can be effectively charged using microwave transmissions having a theoretical beam efficiency of at least 83.7% and an experimental beam efficiency of at least 76.0% [SHI 11b]. This efficiency is sufficiently high to realize the transmission of wireless power using microwaves. For this application, a new GaN Schottky diode was utilized to increase the rectified power and to reduce the EV charging time.

In 2000, a scaled MPT model for running an EV was developed at Kyoto University [SHI 04]. To reduce power loss, the position of the model EV was detected by positioning sensors and microwave power was transmitted only to the position of the model EV.

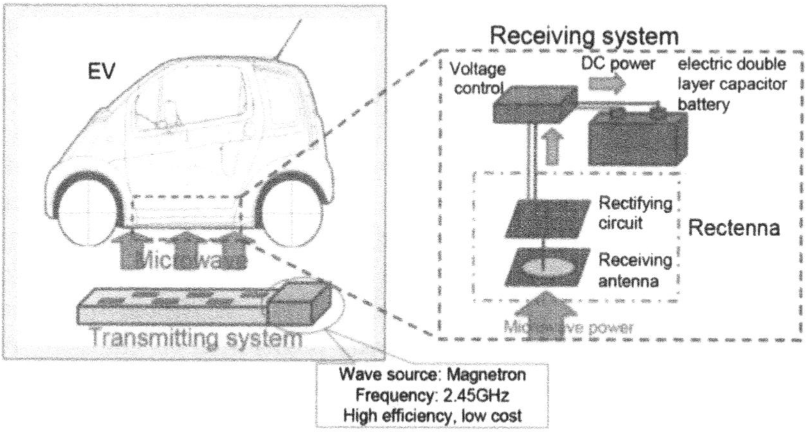

Figure 4.33. *Short-distance wireless microwave charging road system for EVs [SHI 04]*

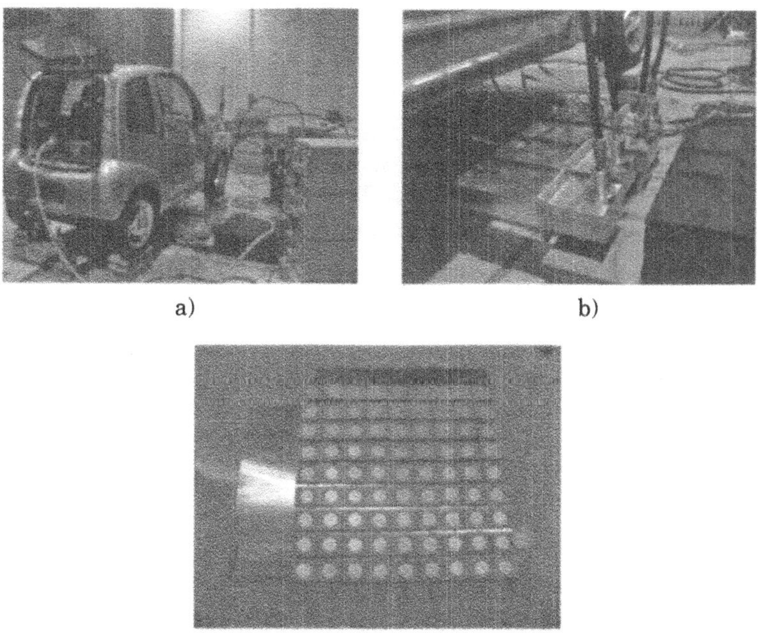

a) b)

c)

Figure 4.34. *Wireless charging experiment with microwaves by Kyoto University: a) the system, b) microwave transmitting antenna on the road and c) rectenna array on the body of the EV [SHI 11b]*

In total, 1–10 kW of microwave power is required to charge the EV. Although the short distance between the transmitting and receiving antennas is an advantage for suppression of unexpected radiation, estimation of the unexpected radiation and a plan to suppress it are essential. Figure 4.35 shows simulation results by the finite difference time domain (FDTD) method of radiation in the system for wireless charging of EVs [SHI 13d]. The radiated microwave was 5 kW at a frequency of 2.45 GHz. A dipole antenna connected to a 50 Ω resistance was positioned on the body of the EV. The body and the roof of the EV were metallic and the windows were unshielded (open). Except between the transmitting and receiving antennas, a safe power density below 1 mW/cm^2 was satisfied.

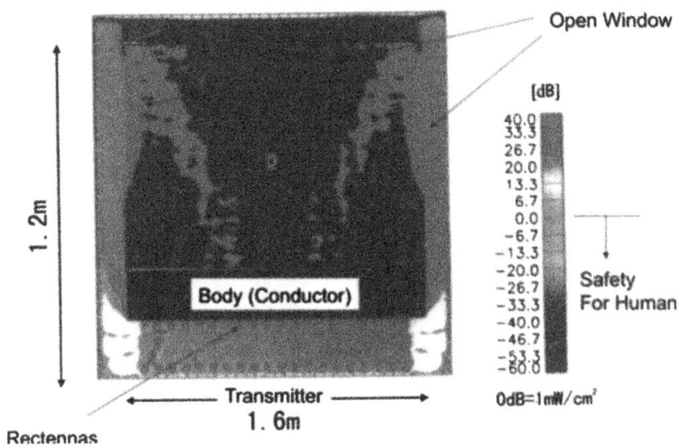

Figure 4.35. *Simulated radiation by the FDTD method of wireless charging of EVs [SHI 13d]. For a color version of this figure, see www.iste.co.uk/shinohara/radiowaves.zip*

From 2006 to 2008, Mitsubishi Heavy Industries Ltd. in Japan conducted an MPT R&D project for EVs with Mitsubishi Motors Co., Fuji Heavy Industries Ltd., and Daihatsu Motor Co., Ltd., at Kyoto University [SHI 13a]. To reduce the power loss, they used: (1) 6.6 kV directly to drive the 2.45 GHz magnetrons as microwave transmitters, (2) a blocking wall around which microwaves pass between the transmitting antennas and receivers and (3) a heat recycling system. The total efficiency, including the heat recycling, was approximately 38% with an output power of 1 kW at a distance of 12.5 cm. The prototype released in 2009 is shown in Figure 4.36.

Figure 4.36. *Wireless charging experiment with microwaves by Mitsubishi Heavy Industries's group in 2009 [SHI 13a]*

In 2012, the Volvo Technologies, Japan group, Nihon Dengyo Kosaku Ltd., and Kyoto University began to develop a new MPT system for an electric track. The former system led to mutual coupling problems between the transmitting and receiving antennas because the MPT distance was too short; therefore, the new MPT system was changed from a road-to-body to a top-to-roof configuration (Figure 4.37) [SHI 13d, SHI 13e] to take advantage of the MPT as long distance WPT. The distance between the transmitting antennas and the receiving antennas on the rooftop of the EV was 2–6 m, depending on the type of EV used. To keep a high efficiency over the varying distance, a phased array system that can create a flat beam on the receiving antennas was proposed. On July 6, 2012, Volvo Technologies, Japan, and

Nihon Dengyo Kosaku Ltd. released a 10 kW rectenna array with an efficiency of 84% operating at a frequency of 2.45 GHz for the mid-distance WPT system (Figure 4.38) [FUR 13]. The received microwave power density was over 3.2 kW/m^2 at a distance of approximately 4 m from the transmitter.

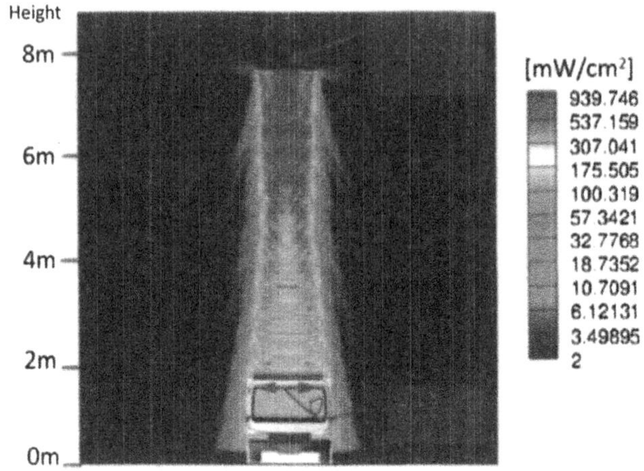

Figure 4.37. *Proposed mid-distance wireless charging for EV and the FDTD simulation of the microwave beam [SHI 13]. For a color version of this figure, see www.iste.co.uk/shinohara/radiowaves.zip*

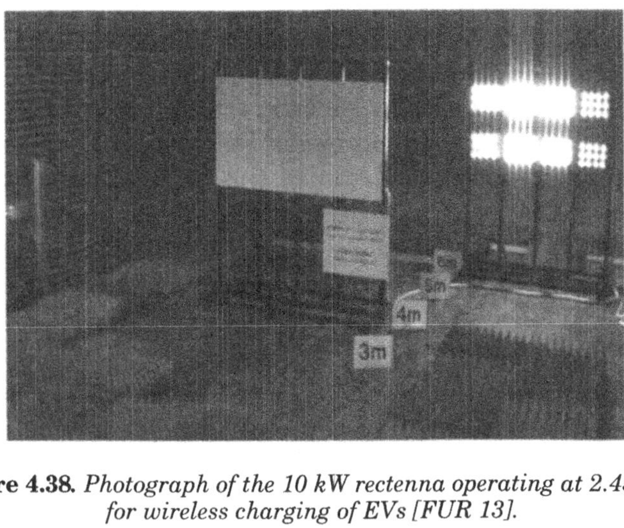

Figure 4.38. *Photograph of the 10 kW rectenna operating at 2.45 GHz for wireless charging of EVs [FUR 13].*

4.9. Point-to-point WPT

It is easy to show a point-to-point WPT via radiowaves in distances of over a kilometer instead of a wired power transfer. In the 1960s, point-to-point WPT operating in the km range via microwaves was highly expected for WPT application. Brown and Dickinson carried out MPT experiments at jet propulsion laboratory (JPL) in 1975 [BRO 84]. However, the size of the transmitting and receiving antennas was decided by theory, and they were too large to realize a commercial point-to-point MPT application with reasonable cost as an alternative to wired power transfer. Point-to-point MPT systems were revised and further experiments were proposed and carried out in the 1990s [SHI 98a, CEL 97]. Under certain conditions, such as providing electrical power to an isolated mountain top or an island where the cost of a wired power supply is too expensive and/or electric power needs are sporadic, the point-to-point MPT system has a distinct advantage over wired power transfer.

The size of the antenna and the expense of an MPT system depends on the MPT's intended range. In Japan, a short-distance point-to-point MPT system has been proposed by Nippon Telegraph and Telephone Corporation (NTT Corp.) and Kyoto University, which is called MPT for fixed wireless access (FWA). Figure 4.39 shows an image of the proposed system [HAT 12]. The outdoor device communicates with the Internet by the FWA or an optical fiber. The inside device and the outside device wirelessly communicate with each other. The inside device transmits power to the outside device by microwave. The outside device can operate without a battery. For the system, it is preferred that both information and power are carried by the same microwave carrier to reduce the size of the system. First, a frequency of 24 GHz was selected and an MMIC rectenna with a class-F load output filter was developed, as shown in Figure 3.83 [HAT 13].

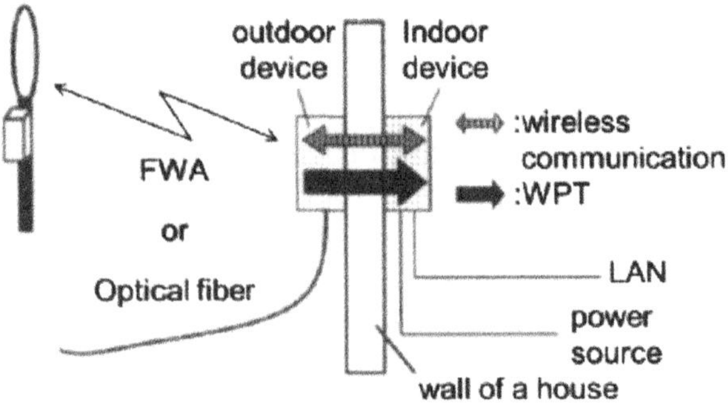

Figure 4.39. *Concept of a microwave-driven FWA system [HAT 12]*

4.10. WPT to moving/flying target

Previous MPT experiments conducted during the SHARP, MILAX and ETHER projects aimed at a stationary high-altitude relay platform in the stratosphere. MPT is the most suitable WPT for a moving/flying target. A phased array is required to control the microwave beam form and direction as was described in section 3.4.

In the early 21st Century, the MPT to microaerial vehicle (MAV) [MIY 12] and Mars observation airplane [NAG 11b, NAG 12] projects, which were MPT systems directed toward small airplanes, were proposed and developed in Japan. The MPT to the MAV was proposed by the University of Tokyo. Researchers transmitted 5.8 GHz microwave power to a flying MAV from which a pilot signal of 2.45 GHz was transmitted as a signal for target detection. Rectennas were installed on the body of the MAV. Initially, five horn antennas were used as a phased array (Figures 4.40 and 4.41). The diameter of the phased array was 330 mm and its element spacing was 2λ. Microwave power from the horn antennas was 4 W each. For the base system, a phased array using eight microstrip antennas whose element spacing was 1.36 λ was adopted. Microwave power from the microstrip antennas was 8 W each.

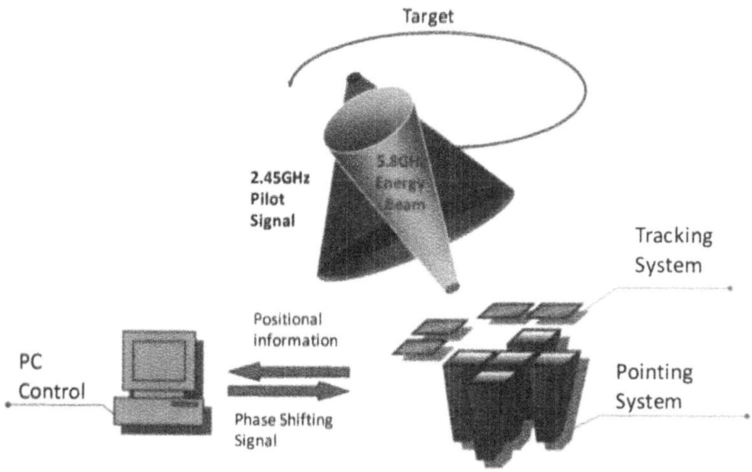

Figure 4.40. *First concept of an MPT to micro-aerial vehicle [MIY 12]*

The Mars observation microwave airplane system was proposed by Kyoto University and Kyusyu Institute of Technology, Japan. Wide-ranging continuous observation of the surface of Mars is of importance for understanding the physical properties of that planet. The surface of Mars has mainly been observed by a rover, which can neither move rapidly nor observe bumpy areas. Therefore, small airplane observation is receiving attention as an alternative to the rover. To realize stable flight in the greatly reduced atmosphere of Mars, the weight of the airplane must be reduced. MPT is an excellent possible technology to reduce or even eliminate the fuel requirements of an airplane. A description of a possible future Mars observation airplane is shown in Figure 4.42 [NAG 11b, NAG 12]. An experimental setup is shown in Figure 4.43. The MPT system using a phased array antenna composed of the "power-variable phase-controlled magnetrons (PVPCMs)" was used in the experiment. The PVPCM is a derivative technology from the phase-controlled magnetron (PCM) developed at Kyoto University. A PVPCM can transmit 2.45 GHz, 61 dBm microwave power and can control the beam direction using phase control [NAG 11b]. The transmitter traces the airplane's location with a camera using an image processing application [NAG 12].

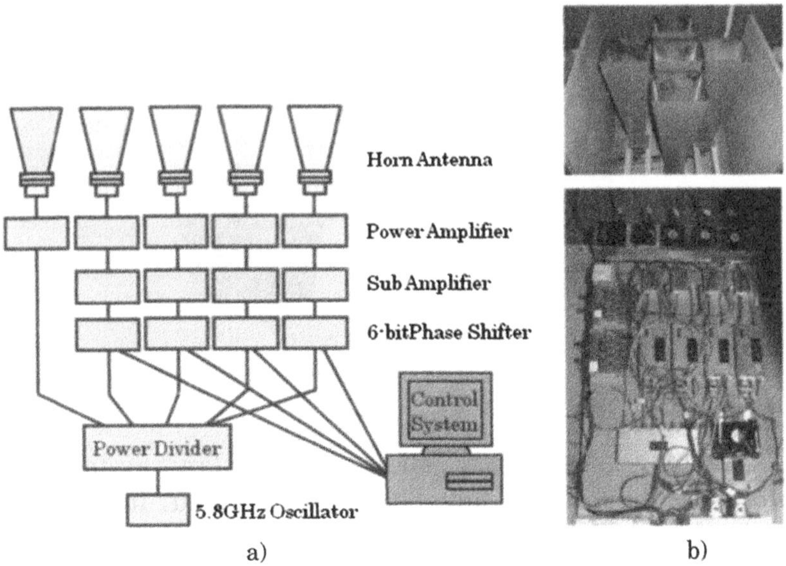

Figure 4.41. *a) Block diagram of a phased array MPT transmitter and b) photograph of the phased array*

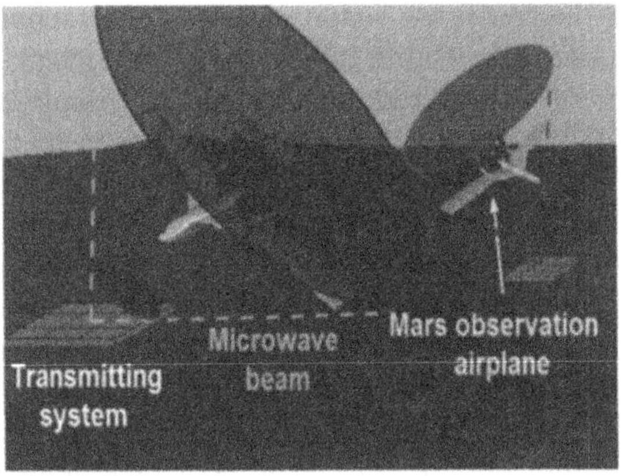

Figure 4.42. *A description of the MPT to Mars observation airplane [NAG 11b, NAG 12]*

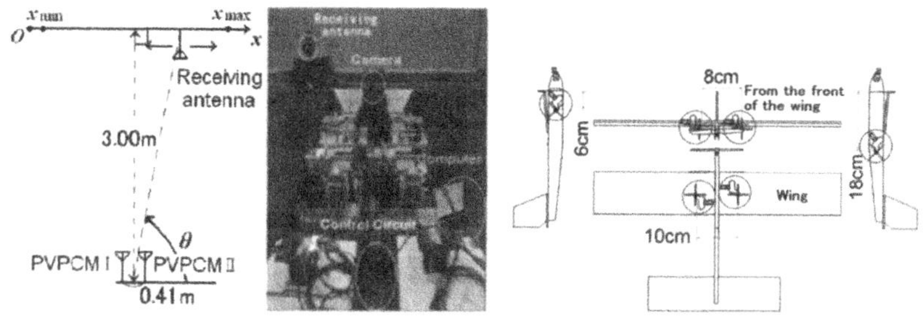

a) b)

Figure 4.43. *a) Schematic diagram and photograph of the magnetron-based phased array and b) alignment of the six rectennas on the airplane body for ground experiments [NAG 12]*

MPT to a moving rover is another potential application. From 2004 in Japan, MPT using active integrated antenna (AIA) technology was developed as a spin-off technology from SPS [SHI 07b]. The target of this project was: (1) development of a very light power-to-weight ratio microwave power transmitter using AIA technology (with a target below 50 g/W), (2) advancement of the power management of the rectified microwave power at the rectenna, receiver and rectifier of the microwave power, especially against changes in the connected load and (3) fundamental experiments of the coexistence of 100 W microwaves and 10 mW wireless communication waves. For the target of the project, MPT to a moving rover was chosen. Figure 4.44 shows the experimental system. The system block diagram is shown in Figure 4.45. The microwave transmitting subsystem was mainly developed at Kyoto University. The microwave transmitting subsystem was composed of a 32 element AIA with a linear polarized rectangular microstrip antenna array and 4W output three-stage GaAs high-power amplifiers on a bent dielectric base for an expanded cooling area, whose total power is 120 W at 5.8 GHz. The system uses no phase shifter. Figure 4.46 shows the developed microwave transmitting subsystem. The total weight of the 120 W system is approximately 4 kg and the power-to-weight ratio is approximately 33.3 g/W. The volume of the subsystem is 0.0544×10^{-3} m^3 ($0.17 \times 0.12 \times 0.32$ m).

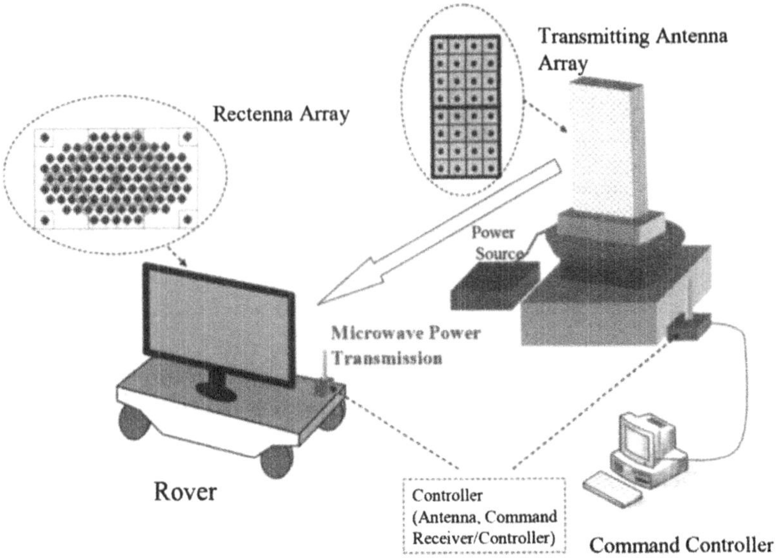

Figure 4.44. *Experimental system of MPT for the moving rover [SHI 07b]*

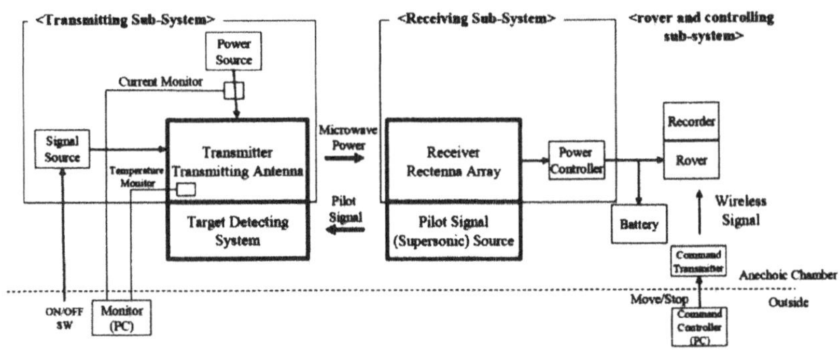

Figure 4.45. *Block diagram [SHI 07b]*

Figure 4.46. *Developed microwave transmitting subsystem with 32 AIA elements and a bent dielectric base for an expanded cooling area [SHI 07b]*

The microwave receiving subsystem was mainly developed at IHI Aerospace. For target detection, supersonic waves and a small turn table were adopted. The microwave beam emitted from the microwave transmitting subsystem positioned on the turn table tracked the rover. A photograph of the MPT experiment in an anechoic chamber is shown in Figure 4.47. The rover moved only by means of the microwave power provided by MPT.

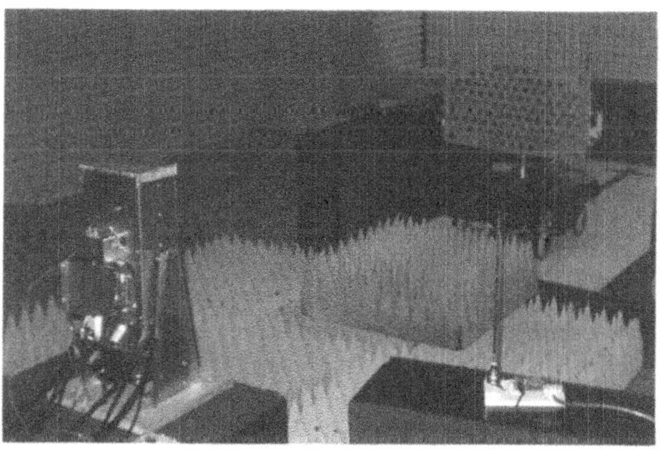

Figure 4.47. *Photograph of the MPT rover experiment [SHI 07b]*

The MPT-rover was applied to an agricultural application. Electrification of agricultural vehicles is one solution to address environmental issues, such as global warming and fossil fuel resource shortages. Lead–acid batteries, however, are heavy and have low power density for agricultural usage. So a microwave-driven agricultural model vehicle was proposed and developed at Kyoto University [OID 07, MIY 13].

The microwave-driven vehicle was built as a model of a small type seeder. The vehicle was equipped with a driving motor (24 V rated voltage, 26 W rated output). A rotary encoder was attached to the drive shaft to measure the vehicle speed. For steering, a DC motor (24 V rated voltage, 14.8 W rated output) was used. The actual steering angle was measured using a rotary potentiometer attached to the steering motor shaft. A rectenna panel was installed on the vehicle as shown in Figure 4.48 [MIY 13].

Figure 4.48. *Microwave-driven agricultural model vehicle [MIY 13]*

The transmitter system is shown in Figure 4.19 [MIY 13]. Microwave power to the vehicle was generated by a magnetron operating at a frequency of 2.45 GHz and transmitted from a parabolic antenna positioned on a turn table. Two monochrome charge coupled device (CCD) cameras mounted on stepper motors were used to capture images of the cylindrical stripe marker mounted on the experimental vehicle. The magnetron output was 800 W, and the maximum total power consumption of the vehicle was approximately 41 W at a distance of a few meters. The vehicle could be solely moved by the microwave power.

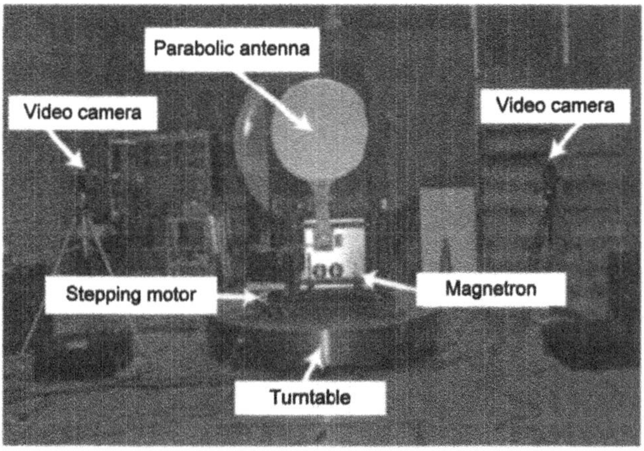

Figure 4.49. *The transmitting system for the microwave-driven agricultural model vehicle [MIY 13]*

4.11. Solar power satellite

4.11.1. *Basic concept*

As previously mentioned, SPSs have been the most suitable application of the MPT since the 1960s because SPSs cannot be realized without MPT. The concept of the SPS is very simple. It is a very large satellite designed as an electric power plant orbiting in a geostationary earth orbit (GEO) 36,000 km above the Earth's surface. An artistic image is shown in Figure 4.50. It consists of mainly three segments: a solar energy collector to convert the solar energy into DC electricity, a DC-to-microwave converter and a large antenna array to beam down the microwave power to the ground. The solar collector can be either photovoltaic cells or solar thermal turbine. The DC-to-microwave converter can be either a microwave tube system and/or a semiconductor system. The third segment is a large phased array. The power beam must be controlled accurately to less than 0.0005°. The SPS system is composed of a space segment and a ground power receiving site. The latter uses a device to receive and rectify the microwave power beam, namely a rectenna array. The rectenna converts the microwave power back to DC power, and is connected to existing electric power networks. The first SPS was proposed by Glaser in 1968 [GLA 68]. After the proposal of the SPS, there were, and continue to be, many SPS research projects worldwide [GLA 93, URS 07, SHI 12b].

Figure 4.50. *Solar power satellite (artist conception)*

The SPS system has an advantage in that it produces electricity with a much higher efficiency compared with a photovoltaic system on the ground. Because the SPS is placed in space in a GEO, there is no atmospheric absorption; therefore, the solar input power has about 30% higher density compared with the ground solar power density, and is available 24 hours a day (except for at most 70 minutes during 42 days near the equinoxes) without being affected by weather conditions. The solar flux is approximately eight times higher in space than the long-term surface average on the ground if the insolation is 4 kWh/m^2/day. However, for terrestrial systems, the efficiency, additional space required to compensate for the losses and cost of the storage system should be taken into consideration in order to supply uninterrupted electricity. This ratio would become higher depending on the efficiency because 100% efficiency is assumed in the storage system. Possible issues to be discussed regarding the SPS system include the effect of the microwave power beam on existing communication networks and organisms.

4.11.2. *SPS as a clean energy source of CO_2-free energy and for a sustainable humanosphere*

The CO_2 emissions per kWh of different electricity generation systems (such as fossil fuels and nuclear power) are compared to SPS, as shown in Table 4.4 [YOS 09]. The CO_2 emitted from the operation of fossil fuel power generation systems is mainly due to the burning of fuel, whereas the CO_2 emissions from nuclear power plants are mainly due to the use of energy to produce the nuclear fuel. Almost zero CO_2 emissions are expected from SPSs during operation. As a result, the SPS system would generate less CO_2 per kilowatt hour than nuclear power generation.

To ensure the sustainability of economic activities, there is a need for the development of a large-scale and clean power source that sufficiently suppresses CO_2 emissions. It is generally accepted that only solar technologies can provide such a large-scale and clean power source in the near future. Terrestrial photovoltaics, wind, geothermal and other natural resources depend on the environmental conditions, and they are neither stable nor sufficient in quantity.

Generating system	Operations	Construction	Total
SPS	0	20	20
Coal	1,222	3	1,225
Oil	844	2	846
LNG	629	2	631
Nuclear power	19	3	22

Table 4.4. *Comparison of relative CO_2 emissions from different electricity generation systems (units: g CO_2 per kWh)*

Considering the Earth as a global system or as "Gaea," the SPS can contribute to a sustainable humanosphere, which includes the Earth and a space system. The Club of Rome, in the "Limits to Growth" on the Earth in 1972 [MEA 72], indicated that the influence of human activities on the Earth, such as the exhaustion of natural resources, pollution of the air, the greenhouse effect, etc., had increased significantly, as previously predicted by Forrester [FOR 73] who performed simulations of world dynamics models. Figure 4.51 shows a basic simulation result of a modified version of Forrester's WORLD-2 model [YAM 92]. For such "Limits to Growth" in the closed Earth system, the SPS will provide the world dynamics model with new boundary conditions and provide new solutions. Figure 4.52 shows a revised simulation result of a world dynamics model incorporating SPSs [YAM 92]. Compared with the result in Figure 4.51, the simulation result shown in Figure 4.52 suggests that the SPS can provide a new solution to remove the limitations to growth. The SPS is one of the potential technologies that can support our life and the Earth, which is called a "sustainable humanosphere" that is open to space systems.

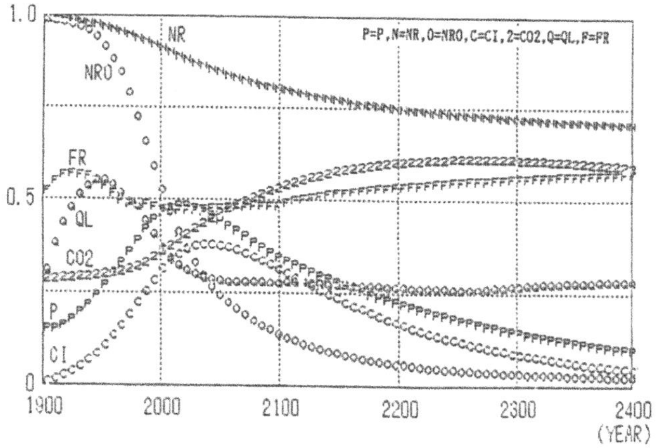

Figure 4.51. *Simulation result of Forrester's WORLD-2 model (modified version). Full-scale values of levels: population (P = 1 × 10^{10} (People)), capital investment (CI = 2 × 10^{10} (capital unit)), total energy resources (NR = 3.24 × 10^{13} (barrels)),oil (NRO = 2 × 10^{12} (barrels)), CO_2 (CO_2 = 1,000 (PPM)), quality of life (QL = 2), food ratio (FR = 2) [YAM 92]*

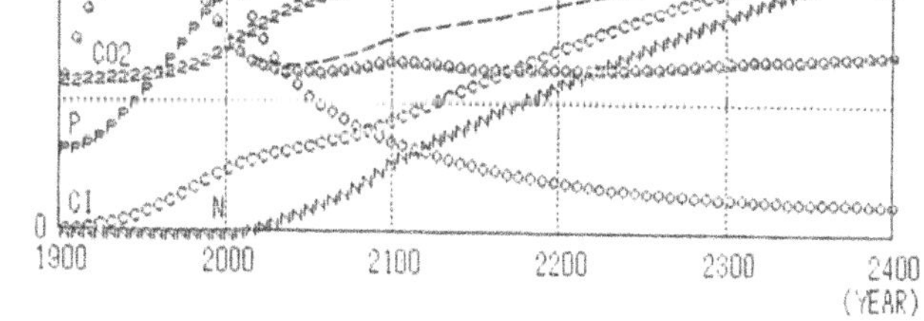

Figure 4.52. *Simulation result of revised world dynamics model with SPS (high-energy investment case). Full-scale values of levels: population (P = 1 × 10^{10} (people)), capital investment (CI = 5 × 10^{10} (capital unit)), SPS number (N = 4,000), oil (NRO = 2 × 10^{12} (barrels)), CO_2 (CO_2 = 1,000 (PPM)), quality of life (QL = 2). Parameters: energy investment in SPS (E_t = 0.003 × NRUR (MJ)), start year of energy investment: (IYEAR = 2000 (Year)), SPS research cost to improve value (RYEAR = 2100), SPS CI discard normal (SCIDN = 0.001) [YAM 92]*

4.11.3. *MPT on SPS*

In the MPT system of the SPS, a large phased array with a high efficiency must be used. The phased array is necessary for steering a power beam to a small rectenna target on the ground within a tolerance of 0.0005°, although the transmitting antenna of the SPS will always move and fluctuate. The power beam must be generated and transmitted without much loss for economic reasons. An economic analysis in Japan indicated that the optimum size of the transmitting phased array should be a few kilometers and the optimum microwave power should be approximately a few GW with a frequency of 2.45 GHz. For the same reason, the DC–RF conversion efficiency (which includes all losses, e.g. in phase shifters, power circuits and isolators) is assumed to be more than 80%. The beam collection efficiency, defined as the ratio of the microwave power received at the rectenna site to the microwave power emitted from the transmitting antenna, which is basically estimated using equation [2.7], is assumed to be approximately 90%. The absorption by the atmosphere is estimated to be below 2%. The weight is also an important parameter of the transmitting antenna for cost estimation. The MPT system, which includes the generator/amplifier, phase shifter and antenna, must be less than several kg/kW to reduce the transportation cost.

Table 4.5 shows some typical parameters of the transmitting antenna of the SPS. An amplitude taper on the transmitting antenna is adopted in order to increase the beam collection efficiency and to decrease the side-lobe levels in almost all SPS designs. A typical 10 dB Gaussian amplitude taper is used, in which the power density in the center of the transmitting antenna is 10 times larger than that on the edge of the transmitting antenna.

Model	NASA/DOE model [DOE 78]	JAXA Model [JAX 05]
Frequency	2.45 GHz	5.8 GHz
Diameter of transmitting antenna	1 kmϕ	1.93 kmϕ
Amplitude taper	10 dB Gaussian	10 dB Gaussian
Output power (beamed to earth)	6.72 GW	1.3 GW
Maximum power density at center	2.2 W/ cm^2	114 mW/cm^2
Minimum power density at edge	0.22 W/ cm^2	11.4 mW/cm^2
Antenna spacing	0.75 λ	0.75 λ
Power per antenna (Number of elements)	Max. 185 W (97 million)	Max. 1.7 W (1,950 million)
Rectenna Diameter	10 km × 13.2 km	2.45 kmϕ
Maximum Power Density	23 mW/cm^2	100 mW/cm^2
Collection Efficiency	89 %	87 %

Table 4.5. *Typical parameters of the transmitting antenna of the SPS*

Details of the microwave beam in the Japan Aerospace Exploration Agency (JAXA) model, which was designed in 2004, are shown in Figure 4.53. Microwave power must remain less than 1 mW/cm^2 outside the rectenna. Compared to the National Aeronautics and Space Administration (NASA)/Department of Energy (DOE) model, which was designed in 1978, the microwave power density on the ground for the JAXA model is greater because Japan is a small country, and the ground system must be smaller.

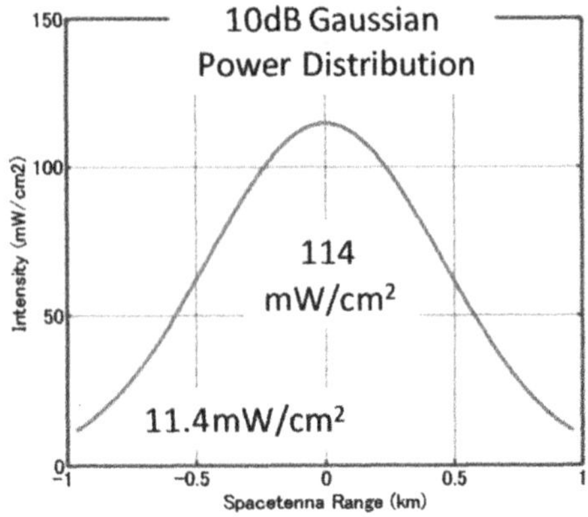

a)

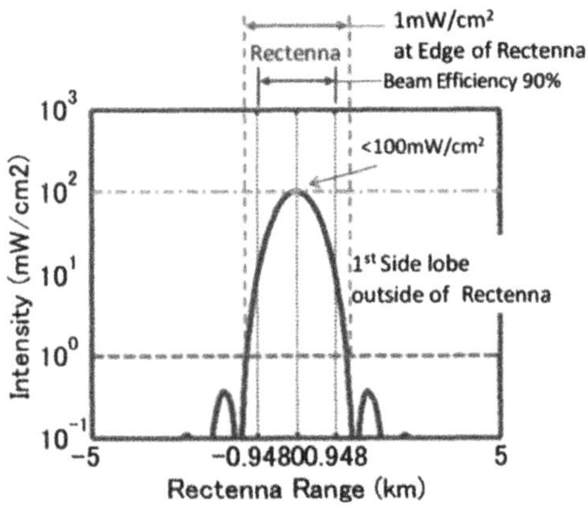

b)

Figure 4.53. *JAXA model: a) power distribution on transmitting antenna whose size is 1.93 km and whose microwave power is 1.3 GW and b) power distribution on rectenna on ground whose size is 2.45 km*

4.11.4. *Various SPS models*

The SPS is the largest and most suitable application of MPT. In Japan, USA and Europe, many studies on SPS have been and are currently being carried out. Figure 4.54 shows committee activities concerning SPS feasibility studies from the 1970s to the present day. The activity had already started in 1977 when the NASA/DOE report was issued. Since then, the committee's activities that were carried out to survey the conceptual design and the feasibility of SPS have continued intermittently up to the present.

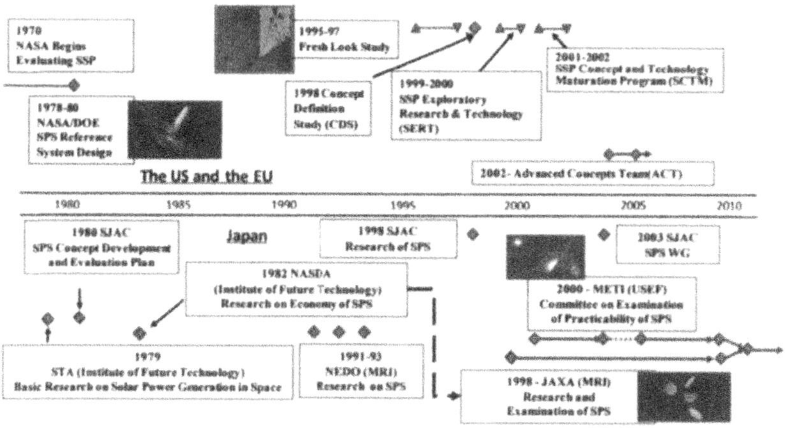

Figure 4.54. *Historical review of global SPS research projects*

4.11.4.1. *SPS in the United States*

Following Glaser's first proposal of the SPS, the United States conducted an extensive feasibility study in 1978–1980. This study was done by a joint effort of NASA and the DOE. They proposed an improved model known as the NASA/DOE reference model in 1979, as shown in Figure 4.55 [DOE 78].

According to this model, a 50 km^2 solar array collects the energy of approximately 70 GW as the sun provides energy of about 1.37 kW/m^2 (137 mW/cm^2) at GEO, and generates 9 GW of DC power (a total efficiency of 13%). This system transmits microwave power of 6.72 GW at 2.45 GHz from a 1 km diameter antenna (78% in conversion efficiency). The number of antenna elements in the SPS would be 100 million in the case of an array antenna with an element spacing of 0.75λ. A 10 dB Gaussian taper is assumed for power distribution in the transmitting antenna in order to obtain a better power collection efficiency. A 10 km diameter rectenna on the ground site at the equator collects 5.8 GW (power collection efficiency of 89%), and 5 GW is sent to the utility grid. The number of rectenna elements is 10 billion with an element spacing of 0.75λ. If the rectenna site is located at a latitude of 35°, its shape should be an ellipse with dimensions 10 km × 13.2 km. Detailed parameters of the MPT are shown in Table 4.5. The overall system efficiency is 7%.

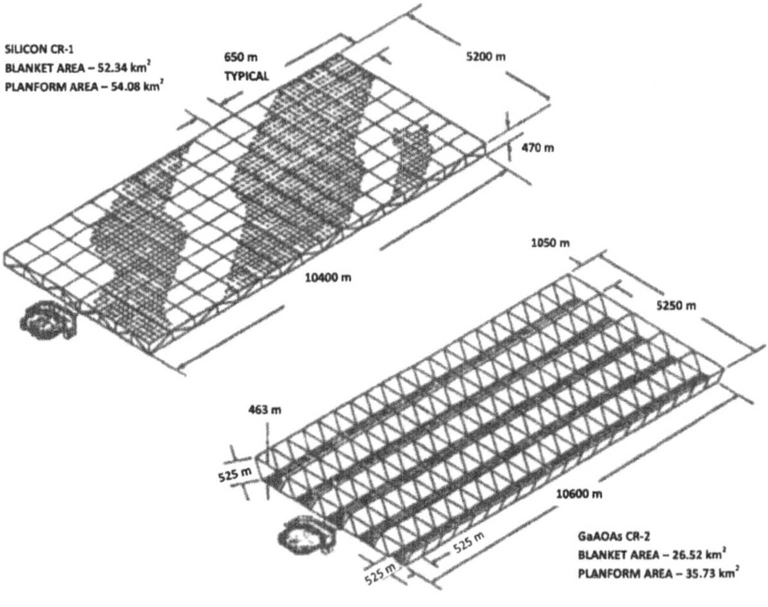

Figure 4.55. *NASA/DOE SPS reference model [DOE 78]*

From the viewpoint of EMC, consideration should be given to ensure the safety of people on the Earth. Although the power density is 23 mW/cm^2 at the center of the array, it is just 1 mW/cm^2 at the edge. The latter satisfies the safety standard, and even the former is a quarter of the solar radiation (100 mW/cm^2). Note that solar radiation is weaker on the ground than in space because of atmospheric absorption.

The architecture of the NASA/DOE reference model entailed using a series of 60 SPSs into GEO. Each of these SPSs was planned to provide a dedicated base-load power ranging from 5 to 10 GW of continuous energy. The platforms were envisioned to be deployed using a massive, unique infrastructure. This infrastructure included a fully reusable two-stage-to-orbit (TSTO) Earth-to-orbit (ETO) transportation system, as well as a massive construction facility in low Earth orbit (LEO). For the construction, hundreds of astronauts would have been required to work continuously in space for several decades. The financial impact of this deployment scheme was significant. In 1996, more than $250 billion was estimated to have been required before the first commercial kilowatt-hour would have been delivered.

Although the SPS research has been suspended in the USA since 1980 because of the high estimated cost at that time, it was not abandoned because of its good potential as a new power source for the next generation. According to the pre-set policy of re-evaluation of the SPS with an appropriate time interval, the Fresh-Look-SPS concepts have been envisioned from 1997 as an improved SPS reference system [MAN 97].

The "Sun Tower" SPS concept is one of their new models that exploits several innovative approaches to reduce the development and lifecycle cost of an SPS, at the same time broadening market flexibility. The concept will entail relatively small individual system components with an extensively evolvable modularity, as depicted in Figure 4.56. Manufacturing can be a "mass production" style from the first satellite system. Therefore, this system can be developed at a moderate price, can be ground-tested with no new facilities and can be demonstrated in a flight environment with a subscale test. This system will initially be deployed in a LEO before subsequent migration to a GEO. This is necessary to achieve extremely low launch costs (of the order of $400 per kg), with payloads of greater than 10 MT; this is consistent with highly reusable space transportation (HRST) system concepts.

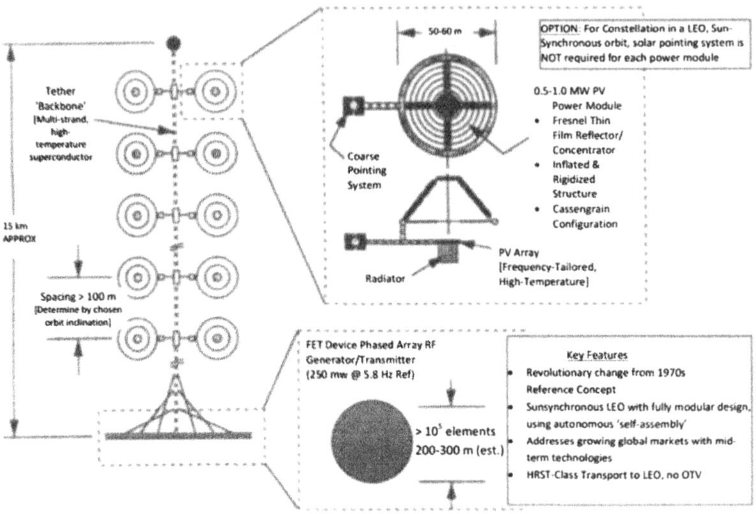

Figure 4.56. *The "Sun Tower" SPS concept (MEO constellation) [MAN 97]*

The "Sun Tower" SPS concept is a constellation of medium-scale, gravity gradient-stabilized satellites. Each satellite resembles a large, Earth-pointing sunflower in which the face of the flower is the transmitter array, and the "leaves" on the stalk are solar collectors. The concept is assumed to transmit 5.8 GHz from an initial operational orbit of 1,000 km, and be sun-synchronous at a transmitted microwave power level of about 200 MW. The total beam-steering capability is 60° (30°). Therefore, a single transmitting element is projected to be a hexagonal surface that is approximately 5 cm in diameter. These elements are pre-integrated into "subassemblies" for final assembly on orbit. For an RF transmitting power of 200 MW, the transmitter array is an "element and subassembly-tiled plane" that is essentially circular, approximately 260 m in diameter and 0.5–1.0 m in thickness.

Sunlight-to-electrical power conversion must be modular and deployable in "units," about 50–100 m in diameter and should have a net electrical output of 1 MW. The primary technology option is a gossamer-structure based on the reflector with non-dynamic conversion at the focus (e.g. advanced solar cells). These sunlight collection systems are presumed to be always sun-facing (with the system in the sun-synchronous orbit) and to be attached regularly in pairs along the length of a structural/power transmitting tether to the backplane of the transmitter array. Heat problems occur both at the surface of the solar cell array and in the transmitter circuit. For both cases, heat rejection for power conversion and conditioning systems is assumed to be modular and integrated with power conversion systems. In the case of the transmitter, the heat rejection is assumed to be both modular and integrated at the "back-plane" of the transmitter array. Single power transmission lines from the central tether attachment point to the backplane are integrated with the modular subassemblies of the array.

The nominal ground receiver for the Sun Tower concept is a 4 km diameter site with a direct electrical feed into the interface with commercial power utilities. The space segment is consistent with a variety of ground segment approaches. However, during the early years of operation, multiple ground stations would be required to achieve reasonable utilization of capacity. For primary power, a ground-based energy storage system would be required, in particular, in the early phases of the overall system deployment in which only a single Sun Tower is operational. A pair of a single satellite and a ground receiver would be sized to approximately 100–400 MW scale, with multiple satellites required to maintain constant power at that level.

After the Fresh-Look Study of the SPS, some SPS research projects were initiated in the United States, which were called the concept definition study (CDS) in 1998, SSP Exploratory Research & Technology (SERT) in 1999–2000 and SSP Concept and Technology Maturation Programs (SCTM) in 2001–2002, respectively [FEI 03]. Through these SPS projects, two new SPS concepts were proposed: Abacus Reflector and Integrated Symmetrical Concentrator (ISC) [FEI 03]. Although SPS was considered in the United States after the SCTM [MAN 09], the ISC is the latest SPS concept present in the USA in 2013.

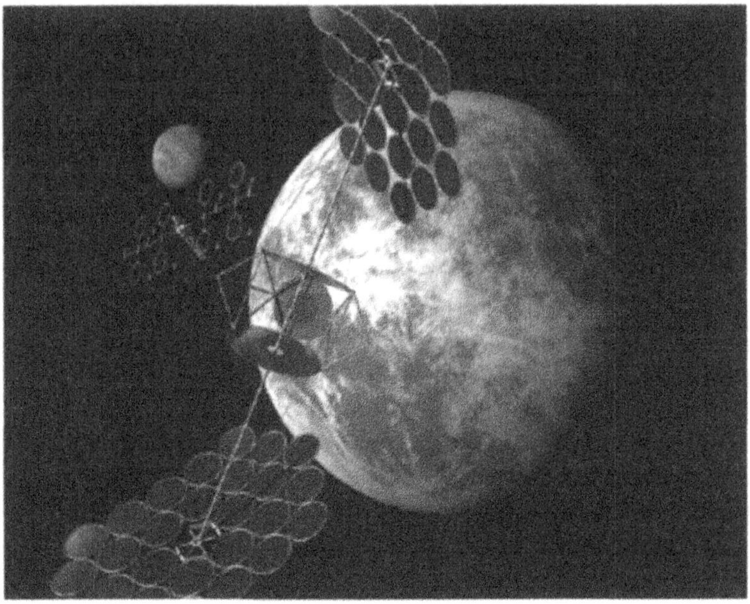

Figure 4.57. *Concept of integrated symmetrical concentrator (ISC) [MAN 09]*

4.11.4.2. *SPS in the EU*

In 2001, the Europeans proposed a Sail Tower SPS, as shown in Figure 4.58 [SEB 01, SEB 04]. The Sail Tower design is similar to NASA's Sun Tower SPS, but uses a thin-film technology and innovative deployment mechanisms developed for solar sails. The main characteristics are summarized in Table 4.6.

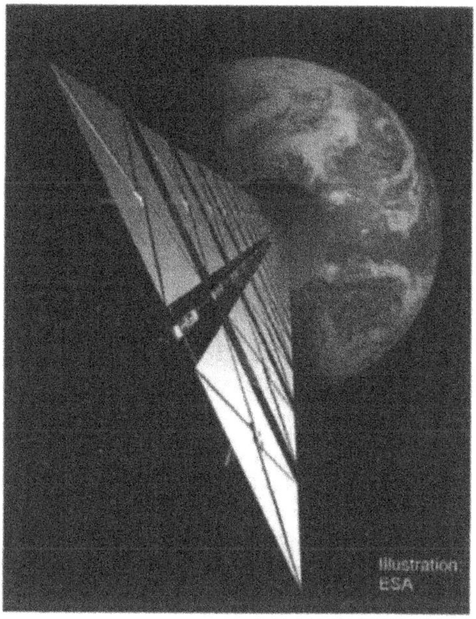

Figure 4.58. *Concept of solar sail [SEB 04]*

Each sail has dimensions 150 m × 150 m and is automatically deployed by extending four lightweight diagonal carbon fiber (CFRP) booms, which are initially rolled up on a central hub. They demonstrated the CFRP booms [LEI 03]. The power generated within the sail modules is transmitted through the central tether to the antenna, where 2.45 GHz microwaves are generated in mass-produced inexpensive magnetrons. Slotted waveguides made with carbon fiber are used as active antenna elements that are mounted on the main antenna structure. For the phased arrays, several sets of waveguides radiate the microwave power to the rectennas on the Earth, where the power is transformed and fed into the existing power distribution networks. The power intensity across the antenna surface is assigned with a truncated 10 dB Gaussian distribution that minimizes side lobes and scatterings.

European Sail Tower SPS			
Orbit	GEO		
Final # of SPS	1,870		
SPS tower	Length	15	[km]
	Mass	2,140	[mt]
	Electricity prod.	450	[MW$_e$]
Twin module	Dim. + tether	150 × 300 × 350	[m]
	Mass	9	[mt]
	Electricity prod.	7.4	[MW$_e$]
Emitting antenna	4,00,000	Magnetron	
	Frequency	2.45	[GHz]
	Radious	510	[m]
	Mass	1,600	[mt]
	Energy emitted	400	[MW]
Receving antenna site	Final number	103	
	Antenna size	11 × 14	[km]
	Site including safety zone	27 × 30	
Power delivered	Per SPS tower	275	[MW$_e$]

Table 4.6. *Characteristics of solar sail*

In 2003, the Advanced Concepts Team (ACT) of the European Space Agency (ESA) began a three-phased, multiyear program related to solar power from space [SUM 03]. In Europe, terrestrial solar power is one of the fastest growing energy sectors with high growth rates that have been sustained over more than a decade, with very promising forecasts. The first phase of the European Program Plan, therefore, involved the terrestrial research community and was dedicated to the assessment of the "general validity" of space concepts for Earth power supply by comparing them with equivalent terrestrial solar concepts. In parallel, the "general validity" of SPS concepts for space exploration and applications were assessed by comparing them with traditional solutions and nuclear power sources.

In January 2005, they summarized the project as follows [ESA 05]:

> Whereas terrestrial solar energy is most economical for supplying power at relatively low (PV) to medium/high load-factors (SOT), space-based solar power is better suited to supplying continuous power and power at high load-factors.

Although the ACT of the ESA has maintained the SPS research activity, its activities have decreased to compared before 2005.

4.11.4.3. *SPS in Japan*

As shown in Figure 4.44, many SPS research projects have been initiated in Japan. From 1991 to 1993, New Energy and Industrial Technology Development Organization (NEDO) conducted an SPS research project in Japan, which was triggered by the then Japanese Prime Minister Toshiki Kaifu, who launched "New Earth 21," which included the SPS. The SPS research committee, whose secretariat was the Mitsubishi Research Institute, Inc., was established in 1991 with 16 members and four Working Groups (WGs). The designed SPS is shown in Figure 4.59 [MIT 94]. The SPS can supply 1 GW on the ground via a 2.45 GHz microwave from the SPS in GEO. First, the SPS is launched to LEO (LEO < 400 km) and is built as shown in the lower right part of Figure 4.59. The initial SPS in the LEO is transported to the GEO as it is, and opens out to its final SPS shape using a centrifugal force. The solar cells are opened on a flexible and light film structure and are supported by control satellites. The weight was estimated at approximately 21 kilotons. Both Si and a-Si were considered as materials for the solar cells on the SPS. The efficiency of the Si was estimated to be approximately 22% with a unit of 10 cm × 10 cm whose thickness was 30 μm. The efficiency of the a-Si was estimated to be 20% with a thickness of 120 μm. The total area of the solar cells was estimated to be 10 km^2, with a weight of 2,300 tons in the case of the Si and area of 11 km^2, and 1,840 tons in the case of the a-Si, respectively. The generated electric power is transmitted to a 30 kVAC, 200 kHz MPT phased array. An inductive coupling was adopted as a rotary joint in the connection between the transmission path and the MPT phased array.

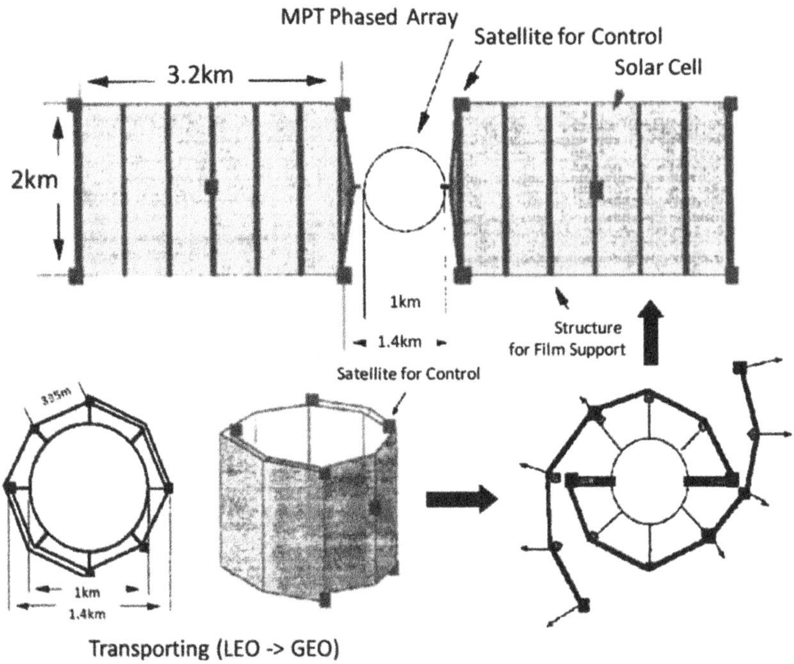

Figure 4.59. *Japanese SPS designed by NEDO SPS committee in 1993 [MIT 94]*

Klystrons were mainly adopted in the MPT system, and the semiconductor system was considered simultaneously. The power distribution on the MPT phase array is Gaussian. The maximum power density at the center is 4.48 kW/m^2 and the minimum at the edge is 448 W/m^2. The antenna element is a dipole with a reflector that can be used as a heat radiator. On the ground, the power density is 23 mW/cm^2 at the center of the rectenna array and less than 1 mW/cm^2 at the edge, respectively. The size of the rectenna on the ground is 10 km × 13 km at sea level.

The Institute of Space and Astronautical Science (ISAS) SPS Working Group, which was formed in 1987, has been engaged in four research projects, one of which is the conceptual design study of a strawman SPS, which was later designated as SPS2000 [NAG 94]. The primary objective of the study was to gain practical knowledge about problem areas of research for SPS systems. A preparatory study was conducted by a special team in the WG for 1 year [NAG 91]. Later, more researchers and groups from the industry and other universities were invited to participate in the first phase design activities. Consequently, 48 members were registered in the task team. The first phase study for the conceptual design was completed in July 1993 [SPS 93]. The design concept of the SPS2000 was to realize a "small and realistic" experimental SPS that is powered from space to developing nations at the equator.

The general configuration of SPS2000 has a shape like that of a triangular prism with a length of 303 m and sides of 336 m, as shown in Figure 4.60. The prism axis is in the latitudinal direction and is perpendicular to the direction of orbital motion. The power transmission antenna, spacetenna, is built on the bottom surface facing the Earth, and the other two surfaces are used to deploy the solar panels.

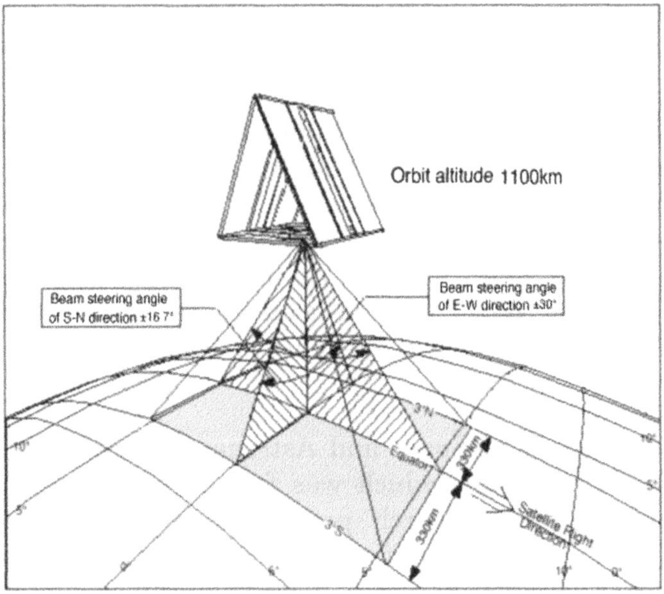

Figure 4.60. *General view of SPS2000 [NAG 94]*

SPS2000 orbits in an equatorial LEO at an altitude of 1,100 km. The orbit is chosen such that it minimizes the transportation cost and the distance of power transmission from space. The spacetenna is constructed as a phased array antenna. It directs a microwave power beam to the position where a pilot signal is transmitted from a ground-based segment of the power system, the rectenna. Therefore, the spacetenna has to be a physically large-size phased array antenna with a retrodirective beam control capability. Therefore, microwave circuits are connected to each antenna element and are driven by DC power generated in the huge solar panels. The frequency of 2.45 GHz is assigned to transmit power to the Earth. The ranges of the beam scan angle are ±30° for the longitudinal direction and ±16.7° for the latitudinal direction. Figure 4.60 also shows a scheme of the microwave beam control and rectenna location. SPS2000 can exclusively serve the equatorial zone, especially benefiting geographically isolated lands in developing nations. The spacetenna has a square shape whose dimensions are 132 m × 132 m, and which is regularly filled with 1,936 subarray segments. The subarray is considered to be a unit of phase control and is a square-shaped array of sides 3 m. It contains 1,320 units of cavity-backed slot antenna elements and DC–RF circuitry. Therefore, there will be approximately 2.6 million antenna elements in the spacetenna.

The JAXA, formerly National Space Development Agency (NASDA), established an SPS research committee in 1998 (until 2007) and studied the SPS's conceptual and technical feasibility at different component levels of the SPS. It is possible to beam the solar energy down to the Earth using either the microwave (radio) technology or the laser (optical) technology. The microwave method in particular has been making rapid progress, whereas the optical methods invariably have weather-related issues. Since 2001, JAXA has proposed a 5.8 GHz, 1 GW SPS model using the microwave technology. Various configurations that are different from the NASA/DOE model have been proposed, evaluated and revised [MOR 04]. In FY2000, the committee developed a prototype of the SPS called SPRITZ (see Figure 3.40).

The JAXA2004 model is the latest JAXA SPS model, which is called "Formation Flying Model," and is shown in Figure 4.61 [JAX 05, ODA 04]. It is based on the formation-flight of a rotating mirror system and an integrated panel composed of a solar cell surface on one side and a phased microwave array antenna on the other side in GEO. The buoyancy can be used to fly the primary mirrors independently. Formation flying mirrors are used to eliminate the need for rotary joints. The whole system becomes mechanically more stable and reliable. The size of the primary mirrors is 2.5 km × 3.5 km (×2 panels), and their weight is 1,000 tons each. The main integrated panel is composed of a solar cell which measures 1.2–2 kmϕ, microwave transmitter and antennas of 1.8–2.5 kmϕ, and two secondary mirrors, whose total weight is approximately 8,000 tons. A 1.31 GW, 5.8 GHz microwave generated with 76% efficiency will be transmitted to a rectenna array on the ground, which measures 2.45 kmϕ. Details of the microwave beam parameters are shown in Table 4.5 and Figure 4.53.

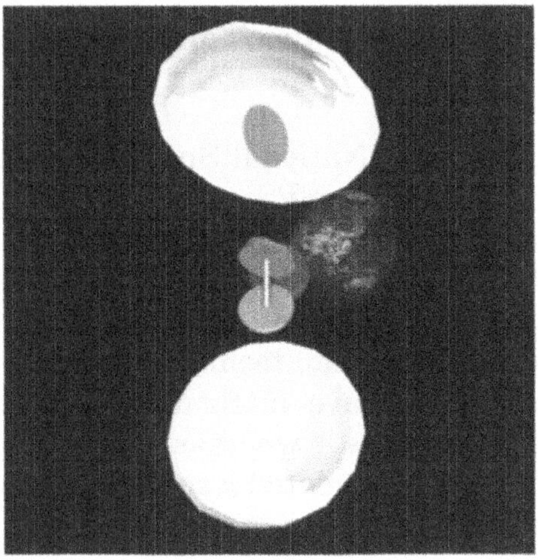

Figure 4.61. *JAXA2004 model of SPS [JAX 05, ODA 04]*

In parallel with the JAXA SPS research committee, there is another SPS research committee that was established by the Ministry of Economy, Trade and Industry (METI) in 2001, the Institute for Unmanned Space Experiment Free Flyer (USEF, present J-Space systems). In the committee, they developed several MPT phased arrays (see Figures 3.45 and 4.44).

They also designed a new SPS which is called tethered-SPS [SAS 04]. An artist's conception of the tethered-SPS is shown in Figure 4.62. The system consists of a large power generation/transmission panel suspended by multitether wires from a bus system above the panel. The attitude is automatically stabilized by the gravity gradient force in the tether configuration without any active attitude control. The power generation/transmission panel consists of perfectly equivalent modules, which significantly contribute to the low-cost production, testing and quality assurance. For the simplest configuration of the power transmission part, the small antenna element and the microwave circuit are integrated as a single entity using an AIA technology. Another innovative feature of the module is the cableless interface realized using a wireless local area network (LAN) system, which leads to reliable deployment, integration and maintenance. The tethered-SPS is an assembly of equivalent miniature tethered-SPS elements. This tethered-SPS is capable of 1.2 GW power transmissions and 0.75 GW average power receptions on the ground. It is composed of a power generation/transmission panel of 2.0 km × 1.9 km suspended by multi-wires deployed from a bus system that is located over 10 km from the power panel. The weights of the panel and the bus system are 18,000 tons and 2,000 tons, respectively. The panel consists of 400 subpanels of 100 m × 95 m with 0.1 m thickness. Each subpanel has 9,500 power generation/transmission modules each with an area of 1 m × 1 m. In each power module, the electric power generated by the solar cells is converted to microwave power and is used for other control units. Therefore, no power line interface exists between the modules. The microwave transmitting antennas are on the lower plane.

Figure 4.62. *Artist conception of the tethered-SPS [SAS 04]*

Because this system has no mechanism to track the sun for the power generation, the total power efficiency is 36% lower than that for the NASA/DOE reference model or other sun-pointing types of SPS. However, it is noted that the simple concept resolves almost all of the technical difficulties in past SPS models. The absence of any moving structure on a large scale makes this system highly robust and stable. Because no light concentrators are used for the power generation, a large-scale kilometer-sized power generation area is required, but the heat generated in the panel can be released into space without the need for any active thermal control.

There is no wired signal interface between the power modules. The control signal and frequency standard for each module are provided from the bus system by the wireless LAN. The tethered subpanel is composed of a 100 m × 95 m subpanel suspended by four wires connected to a bus system. The 100 m × 95 m panel with 0.1 m thickness is regarded as a solid panel with a flatness required for the phase control in the microwave power transmission. The subpanel consists of 950 structural unit panels of 10 m × 1 m × 0.1 m.

The structural unit panels are folded in a package of 9.5 m × 10 m × 10 m, which is a unit cargo transported from the ground to the LEO by reusable launch vehicles (RLV). The cargo is transferred to the orbit transfer vehicle (OTV) in a LEO at approximately 500 km and transported to GEO. To avoid the degradation of the solar cells by trapped energetic particles in the radiation belt, the cargo is housed in a radiation-shielded vessel. If we use a 270 MT OTV equipped with an electric propulsion of 240 N thrust, the cargo would be transferred to GEO within two months. The tethered subpanel is then deployed automatically in GEO. After the function test of the tethered subpanel is completed, it is then integrated into the SPS main body. Docking assistant robots that are controlled by the ground control center will be required for the integration. This strategy makes it possible to verify the SPS function during the construction phase from low-power to full-power transmission.

During the SPS research activities performed by JAXA, ISAS and METI, the Society of Japanese Aerospace Companies (SJAC) sometimes also performed SPS studies. A characteristic of the Japanese SPS research is the presence of continuous and parallel activities.

The Basic Plan for Space Policy in Japan was established by the Strategic Headquarters for Space Policy in June 2009. This Basic Plan for Space Policy that was forged at that time was based on the Basic Space Law, which was established in May 2008, and which is Japan's first basic policy relating to space activities. In the plan, SPS was selected based on nine systems and programs for the use and R&D of space, as follows [BAS 09a, BAS 09b]:

> "As a program that corresponds to the following major social needs and goals for the next 10 years, a Space Solar Power Program will be targeted for the promotion of the 5-year development and utilization plan", and "government will conduct ample studies, then start technology demonstration project in orbit utilizing 'Kibo' (Japanese experimental module on International Space Station) or small sized satellites within the next three years to confirm the influence in the atmosphere and system check".

More than three years have passed from the establishment of the first Basic Plan for Space Policy. During those three years, the Japanese government has changed hands from the Liberal Democratic Party, which had political power for a long time, to the Democratic Party of Japan. Unfortunately, there was less progress in the development of the Space Policy by the Democratic Party of Japan over this three-year period. However, on June 20, 2012, the Japanese government established a new law concerning space development, which is called "Law of partial improvement of the Office of National Space Policy and the Committee on National Space Policy in the Cabinet Office". In the new law, they declared that Cabinet Office will have jurisdiction over Japanese space development rather than several Ministries, including the Ministry of Education, Culture, Sports, Science and Technology, which has jurisdiction over JAXA, which in turn has jurisdiction over the Japanese space development. They also declared that a new space policy decision commission will be established instead of the old Space Activities Commission which had previously led the Japanese space activities. Seven members of the new space policy decision commission were selected in July 26th, 2012 [BAS 12].

In the new Basic Plan for Space Policy, the targets of Japan's space development were narrowed down to the following four systems for the utilization of space and three programs for R&D purposes.

Four social infrastructures for expanding the utilization of space and ensuring autonomy are:

1) Positioning satellites.

2) Remote sensing satellites.

3) Communication/broadcasting satellites.

4) Launching vehicle system.

Three programs that pursue the possibility of future development and utilization of space are:

5) Space Science and Space Exploration Program.

6) Human Space Activity Program.

7) Space Solar Power system R&D Program.

The Space Solar Power or SPS Program was again selected as an R&D program for future space development. In the policy, the R&D target is described as follows:

1) Technology (transportation of large-scale structures to outer space, construction, operation, maintenance, highly efficient and safe power generation, transmission and access to power).

2) Safety (effects on human body, atmosphere, ionized layer, aircrafts, electronic devices, etc.).

3) Economic performance (cutting the cost of transportation to space is the biggest challenge).

The SPS committee formed by METI was restarted from FY2009 based on the Basic Plan for Space Policy in Japan [FUS 11]. The main objective of the restarted committee is the development of a thin and high-efficiency phased array (see Figures 3.48 and 3.49). This activity is jointly supported by METI and JAXA.

Bibliography

[A4W 13] A4WP, http://www.a4wp.org/, 2013.

[ADA 81] ADACHI S., SUZUKI O., ABE S., "Receiving efficiency of an infinite phased array antenna above a reflecting plane", *IEICE Transactions B*, vol. J64-B, no. 6, pp.566–567, 1981. [In Japanese]

[ADL 46] ADLER R., "A study of locking phenomena in oscillators", *Proceedings of the IRE*, vol. 34, pp. 351–357, June 1946.

[AHN 11] AHN S., KIM J., "Magnetic field design for high efficient and low EMF wireless power transfer in on-line electric vehicle", *Proceedings of the EuCAP2011*, pp. 4148–4151, 2011.

[AKI 93] AKIBA R., MIURA K., HINADA M., *et al.*, ISY-METS rocket experiment, The Institute of Space and Astronautical Science Report, no. 652, September 1993.

[AKK 05] AKKERMANS J.A.G., VAN BEURDEN M.C., DOODEMAN G.J.N., *et al.*, "Analytical models for low-power rectenna design", *IEEE Antenna and Wireless Propagation Letters*, vol. 4, pp. 187–190, 2005.

[AOL 13] AOL TECH, http://techcrunch.com/2013/09/09/cota-by-ossia-wireless-power/, 2013.

[BAS 08a] BASIC PLAN FOR SPACE POLICY, http://www.kantei.go.jp/jp/singi/utyuu/basic_plan.pdf, 2008.

[BAS 08b] BASIC PLAN FOR SPACE POLICY Pamphlet, http://www.kantei.go.jp/jp/singi/utyuu/keikaku/pamph_en.pdf, 2008.

[BAS 09a] BASIC PLAN FOR Space POLICY, http://www.kantei.go.jp/jp/singi/utyuu/basic_plan.pdf, 2009.

[BAS 09b] BASIC PLAN FOR SPACE POLICY, http://www.kantei.go.jp/jp/singi/utyuu/keikaku/pamph_en.pdf, 2009.

[BAS 12] BASIC PLAN ON SPACE POLICY, http://www8.cao.go.jp/space/plan/plan-eng.pdf, 2012.

[BOL 78] BOLGER J.G., KIRSTEN F.A., NG L.S., "Inductive power coupling for an electric highway system", *Proceedings of the IEEE 28th Vehicular Technology Conference*, vol. 28, pp. 137–144, March 1978.

[BRO 64] BROWN W.C., "Rectification of microwave power", *IEEE Spectrum*, vol. 1, no. 10, pp. 92–100, October 1964.

[BRO 73a] BROWN W.C., "Adapting microwave techniques to help solve future energy problems", *IEEE Transactions on* Microwave Theory and Techniques, vol. MTT-21, no. 12, pp.753–763, 1973.

[BRO 73b] BROWN W.C., "Adapting microwave techniques to help solve future energy problems", *1973 G-MTT International Microwave Symposium Digest of Technical Papers*, vol. 73, no. 1, pp. 189–191, 1973.

[BRO 76] BROWN W.C., "Optimization of the efficiency and other properties of the rectenna element", *MTT-S International Microwave Symposium*, pp. 142–144, 1976.

[BRO 78] BROWN W.C., "The design of large scale terrestrial rectennas for low-cost production and erection", *Proceedings of the IEEE GMIT International Microwave Symposium*, pp. 349–351, 1978.

[BRO 80] BROWN W.C., "The history of the development of the rectenna", *Proceedings of the SPS Microwave Systems Workshop at JSC-NASA*, pp. 271–280, 1980.

[BRO 81] BROWN W.C., "Status of the microwave power transmission components for solar power satellite", *IEEE Transactions on Microwave Theory and Techniques,* vol. 29, no. 12, pp. 1319–1327, 1981.

[BRO 82] BROWN W.C., TRINER J.F., "Experimental thin-film, etched-circuit rectenna", *MTT-S International Microwave Symposium Digest,* vol. 82, no.1, pp. 185–187, 1982.

[BRO 84] Brown W.C., "The history of power transmission by radio waves", *IEEE Transactions on Microwave Theory and Techniques*, vol. 32, no. 9, pp. 1230–1242, 1984.

[BRO 88] BROWN W.C., "The SPS transmitter designed around the magnetron directional amplifier", *Space Power*, vol. 7, no. 1, pp. 37–49, 1988.

[BRO 89] BROWN W.C., "The Sophisticated properties of the microwave oven magnetron", *IEEE MTTS Digest*, pp. 871–874, 1989.

[BRO 06] BROOKNER E., "Phased arrays and radars – past, present and future", *Proc. of RADAR2002*, pp. 104-113, October 2002.

[BWF 13] BWF, http://bwf-yrp.net/english/, 2013.

[CEL 97] CELESTE A., LUK J.-D.L.S., CHABRIAT J.P., *et al.*, "The Grand-Bassin case study: technical aspects", *Proceedings of the SPS '97*, pp. 255–258, 1997.

[CEL 04] CELESTE A., JEANTY P., PIGNOLET G., "Case study in Reunion island", *Acta Astronautica*, vol. 54, no. 2, pp. 253–258, 2004.

[COL 99] COLANTONIO P., GIANNINI F., LEUZZI G., *et al.*, "On the class-F power amplifier design", *International Journal of RF and Microwave Computer-Aided Engineering*, vol. 9, no. 2, pp. 129–149, March 1999.

[COL 13] COLLADO A., GEORGIADIS A., "24 GHz substrate integrated waveguide (SIW) rectenna for energy harvesting and wireless power transmission", *Proceedings of the International Microwave Symposium (IMS)*, WE4G-3, 2013.

[COV 12] COVIC G., "Recent advances in wireless power transfer for transportation applications", *Proceedings of the 2012 Energy Transfer for Electric Vehicle (ETEV)*, 2012.

[DEN 13] DENGYO CORPORATION, http://www.den-gyo.com/solution/solution10 b.html, 2013. [In Japanese].

[DIA 68] DIAMOND B.L., "A generalized approach to the analysis of infinite planar array antennas", *Proceedings of the IEEE*, vol. 56, pp.1837–1851, 1968.

[DID 98] DIDOMENICO L.D., REBEIZ G.M., "Mobile digital communications using phase conjugating Arrays", *Proceedings of the IEEE Military Communications Conference (MILCOM '98)*, Paper 9.4, 1998.

[DOE 78] DOE AND NASA, Satellite power system; concept development and evaluation program, Reference System Report, October 1978.

[EHC 13] EHC, http://www.keieiken.co.jp/ehc/, 2013. [In Japanese]

[ENE 06] ENERGY HARVESTER, "Method and apparatus for a wireless power supply", Patent No. US7027311, 2006.

[EPP 00] EPP L.W., KHAN A.R., SMITH H.K., *et al.*, "A compact dual polarized 8.51-GHz rectenna for high-voltage (50 V) actuator applications", *IEEE Transactions on Microwave Theory and Techniques*, vol. 48, no. 1, pp. 111–119, 2000.

[ESA 05] ESA, "Earth & Space-Based Power Generation Systems – a comparison study: a study for ESA Advanced Concept Team", Contact No. 17682/03/NL/EC To ESTEC, January 2005.

[FEI 03] FEINGOLD H., CARRINGTON C., "Evaluation and comparison of space solar power concepts", *Acta Astronautica*, vol. 53, nos. 4–10, pp. 547–559, 2003.

[FOR 59] FORRER M.P., Study of multipactor discharge characteristics, RADC Report, No. TN-59-307, Kane Engineering Laboratories, Palo Alto, CA, June 25 1959.

[FOR 73] FORRESTER J.W., *World Dynamics*, 2nd ed., Wright-Allen Press, Inc., Cambridge, 1973.

[FUR 10] FURUKAWA M., SHIRATO T., Efficiency improvement of rectenna for microwave power transmission, IEICE Technical Report, WPT2010-18, pp. 27–30, 2010. [In Japanese]

[FUR 13] FURUKAWA M., MINEGISHI T., OGAWA T., *et al.*, Wireless power transmission to 10kW output 2.4 GHz-band rectenna array for electric trucks application, IEICE Technical Report, WPT2012-47, pp. 36–39, 2013. [In Japanese]

[FUS 11] FUSE Y., SAITO T., MIHARA S., *et al.*, "Outline and progress of the Japanese microwave energy transmission program for SSPS", *Proceedings of the 2011 IEEE MTT-S International Microwave Workshop Series on Innovative Wireless Power Transmission: Technologies, Systems, and Applications (IMWS-IWPT2011)*, pp. 47–50, 2011.

[GAO 13] GAO H., MATTERS-KAMMERER M.K., MILOSEVIC D., *et al.*, "A 62 GHz inductor-peaked rectifier with 7% efficiency", *Proceedings of the 2013 IEEE Radio Frequency Integrated Circuits Symposium (RFIC2013)*, RMO3B-5, 2013.

[GLA 68] GLASER P.E., "Power from the Sun", *Science*, vol. 162, pp. 857–886, 1968.

[GLA 93] GLASER P.E., DAVIDSON F.P., CSIGI K.I. (ed.), *Solar Power Satellite; The Emerging Energy Option*, Ellis Horwood, 1993.

[GOU 61] GOUBAU G., SCHWERING F., "On the guided propagation of electromagnetic wave beams", *IRE Transactions on Antennas and Propagation*, vol. AP-9, pp. 248–256, May 1961.

[GRA 99] GRANATSTEIN V.L., PARKER P.K., ARMSTRONG C.M., "Scanning the technology: vacuum electronics at the dawn of the twenty-first century", *Proceedings of the IEEE*, vol. 87, pp. 702–716, 1999.

[GUO 13] GUO J., ZHU X., "Class F rectifier RF-DC conversion efficiency analysis", *Proceedings of International Microwave Symposium (IMS)*, WE4G-1, 2013.

[GUT 79] GUTMANN R.J., BORREGO J.M., "Power combining in an array of microwave power rectifiers", *IEEE Transactions on Microwave Theory and Techniques*, vol. 27, no. 12, pp. 958–968, 1979.

[HAG 04] HAGERTY J.A., HELMBRECHT F.B., MCCALPIN W.H., *et al.*, "Recycling ambient microwave energy with broad-band rectenna arrays", *IEEE Transactions on MTT*, vol. 52, no. 3, pp. 1014–1024, 2004.

[HAN 11] HANAZAWA M., OHIRA T., "Power transfer for a running automobile", *Proceedings of the 2011 IEEE MTT-S International Microwave Workshop Series on Innovative Wireless Power Transmission: Technologies, Systems, and Applications (IMWS-IWPT2011)*, pp. 77–80, 2011.

[HAR 11] HARAKAWA K., KAGEYAMA K., TSURUTA T., *et al.*, Electrically coupled resonant type wireless power transmission technologies – parallel resonant power transmission circuit, IEICE Technical Report, WPT2011–24, 2011. [In Japanese]

[HAS 11] HASHIMOTO K., ISHIKAWA T., MITANI T., *et al.*, "Improvement of a ubiquitous power source", *Proceedings of the International Union of Radio Science (URSI) General Assembly 2011*, CD-ROM CHGBDJK-1.pdf, 2011. Available at http://www.ursi.org/en/home.asp.

[HAT 98] HATFIELD M.C., HAWKINS J.G., BROWN W.C., "Use of a magnetron as a high-gain, phase-locked amplifier in an electrically-steerable phased array for wireless power transmission", *Proceedings of the 1998 MTT-S International Microwave Symposium*, pp. 1157–1160, 1998.

[HAT 99] HATFIELD M.C., HAWKINS J.G., "Design of an electronically-steerable phased array for wireless power transmission using a magnetron directional amplifier", *Proceedings of the 1999 MTT-S International Microwave Symposium*, pp. 341–344, 1999.

[HAT 12] HATANO K., SHINOHARA N., MITANI T., *et al.*, "Development of improved 24GHz-band class-F load rectennas", *Proceedings of the 2012 IEEE MTT-S International Microwave Workshop Series on Innovative Wireless Power Transmission: Technologies, Systems, and Applications (IMWS-IWPT2012)*, pp. 163–166, 2012.

[HAT 13] HATANO K., SHINOHARA N., SEKI T., *et al.*, "Development of MMIC Rectenna at 24GHz", *Proceedings of the 2013 IEEE Radio & Wireless Symposium (RWS)*, pp. 199–201, 2013.

[HEI 97] HEIDER S., "The commercial space TWTA market review and trends", *Proceedings of the 1997 ESA Workshop*, pp. 63–68, 1997.

[HIN 13] HINO MOTORS, http://hino.dga.jp/i-viewer_s/?p_no=7&m_p=20&p_id =1983& file_name=http %3A%2F%2Fwww.hino-global.com%2Fpdf% 2Fhinorep2007_ e.pdf&t=HINO+Report&kw= IPT+hybrid, 2013.

[HIR 97] HIRAYAMA K., SHINOHARA N., HASHIMOTO K., *et al.*, "Fundamental study of microwave power transmission to a robot moving in gas pipes", *Proceedings of IEICE Comm.*, p. 116, March 1997. Available at http://www.ieice.org/eng/index.html. [In Japanese]

[HIR 99] HIRAYAMA K., SHINOHARA N., MATSUMOTO H., *et al.*, "Study of microwave power transmission to a robot moving in gas pipes", *Proceedings of IEICE Comm.*, p. 25, March 1999. Available at http://www.ieice.org/eng/index.html. [In Japanese]

[HOM 11] HOMMA Y., SASAKI T., NAMURA K., *et al.*, "New phased array and rectenna array systems for microwave power transmission research", *Proceedings of the 2011 IEEE MTT-S International Microwave Workshop Series on Innovative Wireless Power Transmission: Technologies, Systems, and Applications (IMWS-IWPT2011)*, pp. 59–62, 2011.

[HON 11] HONJO K., "High power amplifier for MPT", in SHINOHARA N. (ed.), *Solar Power Satellite, Ohm Publishing*, pp. 50–62, 2011. [In Japanese]

[HUT 94] HUTIN M., LE-BLANC M., Transformer system for electric Railways, US Patent Number 527,875, 1894.

[ICH 12] ICHIHARA T., MITANI T., SHINOHARA N., "Study on intermittent microwave power transmission to a ZigBee device", *Proceedings of the 2012 IEEE MTT-S International Microwave Workshop Series on Innovative Wireless Power Transmission: Technologies, Systems, and Applications (IMWS-IWPT2012)*, pp. 209–212, 2012.

[ICN 13] ICNIRP, http://www.icnirp.de/, 2013.

[IKE 02] IKEMATSU H., MIZUNO T., SATOH H., *et al.*, "SPS concept with high efficiency phase control technology", *Proceedings of the Asia-Pacific Microwave Conference 2002*, WS-9-2, 2002.

[IMU 09] IMURA T., UCHIDA T., HORI Y., "Basic experimental study on helical antennas of wireless power transfer for electric vehicles by using magnetic resonant couplings", *Proceedings of the IEEE Vehicle Power and Propulsion Conference*, pp. 936–940, 2009.

[IMU 10] IMURA T., UCHIDA T., HORI Y., "Flexibility of contactless power transfer using magnetic resonance coupling to air gap and misalignment for EV", *World Electric Vehicle Association Journal*, vol. 3, pp. 1–10, 2010.

[ITO 84] ITOH K., OHGANE T., OGAWA Y., "Absorption efficiency of rectenna composed of magnetic current antennas", *Technical Report of IEICE*, AP84-69, pp. 9–14, 1984.

[ITO 86] ITOH K., OHGANE T., OGAWA Y., "Rectenna composed of a circular microstrip antenna", *Space Power*, vol. 6, pp. 193–198, 1986.

[ITO 93] ITOH T., FUJINO Y., FUJITA M., "Fundamental experiment of a rectenna array for microwave power reception", *IEICE Transactions on Communications,* vol. E76-B, no. 12, pp. 1508–1513, 1993.

[ITU 00] ITU Radiocommunication Study Group, "Applications and characteristics of wireless power transmission", Document No. 1A/18-E, Task Group ITU-R WP1A, Reference Question 210/1, 9 October 2000.

[JAX 05] JAXA, "SSPS research report", March 2005. [*In Japanese*]

[KAM 05] KAMO Y., *et al.*, "C-band 140 W AlGaN/GaN HEMT with Cat-CVD technique", *Proceedings of the 2005 IEEE MTT-S International Microwave Symposium,* WE1E-4, June 2005.

[KAM 11] KAMIYAMA M., ISHIKAWA R., HONJO K., "C-band high efficiency AlGaN/GaN HEMT power amplifier by controlling phase angle of harmonics", *Proceedings of the IEICE*, CS-3-1, September 2011. [In Japanese]

[KAW 13] KAWAHARA Y., WEI W., NARUSUE Y., *et al.*, "Virtualizing power cords by wireless power transmission and energy harvesting", *Proceedings of the 2013 IEEE Radio & Wireless Symposium (RWS)*, pp. 37–39, 2013.

[KAY 86] KAYA N., MATSUMOTO H., MIYATAKE S., *et al.*, "Nonlinear interaction of strong microwave beam with the ionosphere: MINIX rocket experiment", *Space Solar Power Review*, vol. 6, pp. 181–186, 1986.

[KAY 93] KAYA N., MATSUMOTO H., AKIBA R., "Rocket experiment METS microwave energy transmission in space", *Space Power*, vol. 11, nos. 1–2, pp. 267–274, 1993.

[KAY 96] KAYA N., IDA S., FUJINO Y., *et al.*, "Transmitting antenna system for airship demonstration (ETHER)", *Space Energy and Transportation*, vol. 1, no. 4, pp. 237–245, 1996.

[KAY 06] KAYA N., IWASHITA M., TANAKA K., *et al.*, "Rocket experiment on microwave power transmission with Furoshiki deployment", *Proceedings of the International Astronautical Congress (IAC) 2006*, IAC-06-C3.3.03.pdf, 2006.

[KIM 02] KIM S., SHINOHARA N., MATSUMOTO H., "Development of a stable RF-DC conversion efficiency of rectenna on various loads", *Proceedings of the 5th SPS Symposium*, pp. 83–88, 2002. [In Japanese]

[KIM 03] KIM S., SHINOHARA N., MATSUMOTO H., "Development of high conversion efficiency method of rectenna system with reflection", *Proceedings of the IEICE in Spring*, p.C-2-97, 2003. [In Japanese]

[KIM 04] KIMURA T., YAMAMOTO K., NAKADA T., USEF SSPS STUDY TEAM, "Development of highly efficient active integrated antenna", *Proceedings of the 4th International Conference on Solar Power from Space (SPS '04)*, pp. 125–130, 2004.

[KIT 06] KITAYOSHI H., SAWAYA K., "A study on rectenna for passive RFID-tag", *Proceedings of the IEICE in Spring*, CBS1-5, 2006. [In Japanese]

[KIT 13] KITAZAWA S., HANAZAWA M., ANO S., *et al.*, "Field test results of RF energy harvesting from cellular base station", *Proceedings of the 6th Global Symposium on Millimeter-Waves (GSMM) 2013*, T6-6, 2013. Available at http://www.katolab.riec.tohoku.ac.jp/gsmm2013/.

[KOB 93] KOBAYASHI Y., SEKI H., ITOH M., "Improvement of a rectifier circuit of rectenna element for the stratosphere stationary radio relay system", *Proceedings of the IEICE in Spring*, pp. 2–37, 1993. [In Japanese]

[KUR 07] KURS A., KARALIS A., MOFFATT R., *et al.*, "Wireless power transfer via strongly coupled magnetic resonances", *Science*, vol. 317, pp. 83–86, 2007.

[KWP 13] KWPF, http://www.kwpf.org/eng/html/main.html, 2013.

[LAD 13] LADAN S., WU K., "High efficiency low-power microwave rectifier for wireless energy harvesting", *Proceedings of the International Microwave Symposium (IMS)*, WE4G-2, 2013.

[LEI 03] LEIPOLD M., EIDEN M., GARNER C.E., *et al.*, "Solar sail technology development and demonstration", *Acta Astronautica*, vol. 52, nos. 2–6, pp. 317–326, 2003.

[LIP 87] LIPSKY S.E., *Microwave Passive Direction Finding*, Wiley-Interscience, 1987.

[MAE 05] MAEKAWA A., *et al.*, "A 100W high-efficiency GaN HEMT amplifier for S-band wireless system", *Proceedings of the 2005 European Microwave Conference*, pp. 497–500, October 2005.

[MAE 06] MAEKAWA A., *et al.*, "A 500W push-pull AlGaN/GaN HEMT amplifier for L-band high power application", *Proceedings of the 2006 IEEE MTT-S International Microwave Symposium*, pp. 722–725, June 2006.

[MAE 13] MAEHARA D., AKAI R., TRAN G.K., *et al.*, "Experiment on battery-less sensor activation via multi-point wireless energy transmission", *Proceedings of the 2013 IEEE 24th Annual International Symposium on Personal, Indoor, and Mobile Radio Communications (PIMRC)*, pp. 2346–2350, 2013.

[MAI 94] MAILLOUX R.J., *Phased Array Antenna Handbook*, Artech House, 1994.

[MAN 97] MANKINS J.C., "A fresh look at space solar power: new architectures, concepts and technologies", *Acta Astronautica*, vol. 41, nos. 4–10, pp. 347–359, 1997.

[MAN 09] MANKINS J.C., "New detections for space solar power", *Acta Astronautica*, vol. 65, nos. 1–2, pp. 146–156, 2009.

[MAT 86] MATSUMOTO H., KIMURA T., "Nonlinear excitation of electron cyclotron waves by a monochromatic strong microwave: computer simulation analysis of the MINIX results", *Space Solar Power Review*, vol. 6, pp. 187–191, 1986.

[MAT 93] MATSUMOTO H., KAYA N., FUJITA M., *et al.*, "MILAX Airplane experiment and model airplane", *Proceedings of the 11th ISAS Space Energy Symposium*, pp. 47–52, 1993. [In Japanese]

[MAT 95a] MATSUMOTO H., "Microwave power transmission from space and related nonlinear plasma effects", *The Radio Science Bulletin*, vol. 273, pp. 11–35, 1995.

[MAT 95b] MATSUMOTO H., HIRATA H., HASHINO Y., *et al.*, "Theoretical analysis of nonlinear interaction of intense electromagnetic wave and plasma waves in the Ionosphere", *Electronics and Communications in Japan*, Part 3, vol. 78, no. 11, pp. 104–11, 1995.

[MAT 95c] MATSUMOTO H., HASHINO Y., YASHIRO H., SHINOHARA N., OMURA Y., "Computer simulation on nonlinear interaction of intense microwave with space plasmas", *Electronics and Communications in Japan*, Part 3, vol. 78, no.11, pp. 89–103, 1995.

[MAT 02a] MATSUMOTO H., "Research on solar power satellite and microwave power transmission in Japan: review and perspectives", *IEEE Microwave Magazine*, vol. 3, no.4, pp. 36–45, December 2002.

[MAT 02b] MATSUMOTO H., SHINOHARA N., HASHIMOTO K., Activities of study of solar power satellite/station (SPS) in RASC of Kyoto University, Technical Report of IEICE, SPS2002-07 (2002-11), pp. 9–14, 2002. [In Japanese]

[MCS 96] MCSPADDEN J.O., CHANG K., PATTON A.D., "Microwave power transmission research at Texas A&M University", *Space Energy and Transportation,* vol. 1, no. 4, pp. 368–393, 1996.

[MCS 98] MCSPADDEN J.O., FAN L., CHANG K., "Design and experiments of a high- conversion-efficiency 5.8-GHz rectenna", *IEEE Transactions on Microwave Theory and Techniques*, vol. 46, no. 12, pp. 2053–2060, 1998.

[MCS 02] MCSPADDEN J.O., MANKINS J.C., "Space solar power programs and microwave wireless power transmission technology", *IEEE Microwave Magazine*, vol. 3, no. 4, pp. 46–57, December 2002.

[MEA 72] MEADOWS D.H., MEADOWS D.L., RANDERS J., *et al.*, *The Limits to Growth*, Universe Books, New York, 1972.

[MEI 12] MEINS I.J., "Basic applications of inductive power transfer", *Proceedings of the 2012 Energy Transfer for Electric Vehicle (ETEV)*, 2012. Available at http://edpc.eu/etev/etev.html.

[MIK 04] MIKAMI I., MIZUNO T., IKEMATSU H., *et al.*, "Some proposals for the SSPS actualization from innovative component technology standpoint", *Proceedings of the 2004 URSI EMT-S*, pp. 317–319, 2004.

[MIT 94] MITSUBISHI RESEARCH INSTITUTE, INC., Final report of SPS research trusted from NEDO, 1994. [In Japanese]

[MIT 03] MITANI T., SHINOHARA N., MATSUMOTO H., *et al.*, "Experimental study on oscillation characteristics of magnetron after turning off filament current", *Electronics and Communications in Japan, Part II: Electronics*, vol. E86, no. 5, pp. 1–9, 2003.

[MIT 10] MITANI T., YAMAKAWA H., SHINOHARA N., *et al.*, "Demonstration experiment of microwave power and information transmission from an Airship", *Proceedings of the 2nd International Symposium on Radio System and Space Plasma*, pp. 157–160, 2010.

[MIU 01] MIURA T., SHINOHARA N., MATSUMOTO H., "Experimental study of rectenna connection for microwave power transmission", *Electronics and Communications in Japan*, Part 2, vol. 84, no. 2, pp. 27–36, 2001.

[MIY 01] MIYAMOTO R.Y., QIAN Y., ITOH T., "An active integrated retrodirective transponder for remote information retrieval-on-demand", *IEEE Transactions on MTT*, vol. 49, no. 9, pp. 1658–1662, 2001.

[MIY 02] MIYAMOTO R.Y., ITOH T., "Retrodirective arrays for wireless communications", *IEEE Microwave Magazine*, vol. 3, no. 1, pp. 71–79, March 2002

[MIY 12] MIYASHIRO K., INOUE F., MAKI K., *et al.*, Sequentially rotated array antenna for wireless power transmission to an MAV, IEICE Technical Report, WPT2012-30, pp. 59–61, 2012. [In Japanese]

[MIY 13] MIYASAKA J., OHDOI K., WATANABE M., *et al.*, "Control for microwave-driven agricultural vehicle – tracking system of parabolic transmitting antenna and vehicle rectenna panel", *Engineering in Agriculture, Environment and Food (EAEF)*, vol. 6, no. 3, pp. 135–140, 2013.

[MIZ 04] MIZUNO T., NISHIDA K., OKEGAWA H., *et al.*, "Development of a PLL-heterodyne hardware retro directive antenna", *Proceedings of the 48th Aerospace Science and Technology Conference*, 1B06, 2004. [In Japanese]

[MOR 04] MORI M., NAGAYAMA H., SAITO Y., *et al.*, "Summary of studies on space solar power systems of the national space development agency of Japan", *Acta Astronautica*, vol. 54, no. 5, pp. 337–345, 2004.

[MOR 06] MORI K., KATAYAMA A., TSUTUSMI K., *et al.*, "UHF band RF circuits for RFID tag with battery supply", *Proceedings of The Institute of Electronics, Information and Communication Engineers (IEICE) in Spring*, CBS1-6, 2006. Available at http://www.ieice.org/eng/index.html. [In Japanese]

[NAG 86] NAGATOMO M., KAYA N., MATSUMOTO H., "Engineering aspect of the microwave-ionosphere nonlinear interaction experiment (MINIX) with a sounding rocket", *Acta Astronautica*, vol. 13, pp. 23–29, 1986.

[NAG 91] NAGATOMO M., ITOH K., "An evolutionary satellite power system for international demonstration in developing nations", *Proceedings of SPS '91- Power from Space*, 1991.

[NAG 94] NAGATOMO M., SASAKI S., NARUO Y., "Conceptual study of a solar power satellite, SPS 2000", *Proceedings of the 19th International Symposium on Space Technology and Science (ISTS1994)*, pp. 469–476, May 1994.

[NAG 11] NAGAHAMA A., MITANI T., SHINOHARA N., *et al.*, "Study on a microwave power transmitting system for mars observation airplane", *Proceedings of the 2011 IEEE MTT-S International Microwave Workshop Series on Innovative Wireless Power Technologies, Systems, and Applications (IMWSIWPT2011)*, pp. 63–66, 2011.

[NAG 12] NAGAHAMA A., MITANI T., SHINOHARA N., *et al.*, "Auto tracking and power control experiments of a magnetron-based phased array power transmitting system for a mars observation airplane", *Proceedings of the 2012 IEEE MTT-S International Microwave Workshop Series on Innovative Wireless Power Transmission: Technologies, Systems, and Applications (IMWS-IWPT2012)*, pp. 29–32, 2012.

[NAK 05] NAKASUKA S., FUNANE T., NAKAMURA Y., *et al.*, "Sounding rocket flight experiment for demonstrating 'FUROSHIKI SATELLITE' for large phased array antenna", *Proceedings of the International Astronautical Congress (IAC) 2005*, IAC-05-C3.3.01.pdf, 2005.

[NAK 06] NAKASUKA S., FUNANE T., NAKAMURA Y., *et al.*, "Sounding rocket flight experiment for demonstrating 'FUROSHIKI SATELLITE' for large phased array antenna", Acta Astronautica, vol. 59, no. 1–5, pp.200–205, 2006. Available at http://www.sciencedirect.com/science/article/pii/S0094576506001056.

[NOD 11] NODA A., SHINODA H., "Selective wireless power transmission through high-Q flat waveguide-ring resonator on 2-D waveguide sheet", *IEEE Transactions on MTT*, vol. 59, no. 8, pp. 2158–2167, 2011.

[NOD 12a] NODA A., SHINODA H., "Compact class-F RF-DC converter with antisymmetric dual-diode configuration", *Proceedings of the International Microwave Symposium (IMS)*, TH1A-5, 2012.

[NOD 12b] NODA A., SHINODA H., "Waveguide-ring resonator coupler with class-F rectifier for 2-D waveguide power transmission", *Proceedings of the 2012 IEEE MTT-S International Microwave Workshop Series on Innovative Wireless Power Transmission: Technologies, Systems, and Applications (IMWS-IWPT2012)*, pp. 259–262, 2012.

[NOD 13] NODA A., SHINODA H., "A phased array feeding system for 2-D waveguide power transmission", *Proceedings of IEICE*, BSC-1-8, March 2013. [In Japanese]

[ODA 04] ODA M., "Realization of the solar power satellite using the formation flying solar reflector", *Proceedings of NASA Formation Flying Symposium*, September 2004.

[OHI 13] OHIRA T., "Power efficiency and optimum load formulas on RF rectifiers featuring flow-angle equations", *IEICE Electronics Express*, vol. 10, no. 11, pp. 1–9, 2013.

[OID 07] OIDA A., NAKASHIMA H., MIYASAKA J., *et al.*, "Development of a new type of electric off-road vehicle powered by microwaves transmitted through air", *Journal of Terramechanics*, vol. 44, no. 5, pp. 579–587, 2007

[OKA 06] OKAMOTO Y., *et al.*, "100W C-band single-chip GaN FET power amplifier", *Electronics Letters*, vol. 42, no. 5, pp. 283–285, March 2006.

[OTS 90] OTSUKA M., OMURO N., KAKIZAKI K., *et al.*, "Relation between spacing and receiving efficiency of finite rectenna array", *IEICE Transactions B-II*, vol. J74-B-II, no. 3, pp. 133–139, 1990. [In Japanese]

[OTT 74] OTTO D., New Zealand Patent No. 167,422, 1974.

[PAL 01] PALMOUR J.W., *et al.*, "Wide bandgap semiconductor devices and MMICs for RF power applications", *Proceedings of the 2001 International Electron Device Meeting*, pp. 17.4.1–17.4.4, December 2001.

[PHA 13] PHAM B.L., PHAM A.-V., "Triple bands antenna and high efficiency rectifier design for RF energy harvesting at 900, 1900 and 2400 MHz", *Proceedings of the International Microwave Symposium (IMS)*, WE3G-5, 2013.

[PLU 13] PLUGLESS POWER, http://www.pluglesspower.com/, 2013.

[PMA 13] PMA, http://www.powermatters.org/, 2013.

[POW 13a] POWERCAST, http://www.powercastco.com/products/power harvester-receivers/, 2013.

[POW 13b] POWERCAST CORPORATION, http://www.powercastco.com/, 2013.

[QIA 98] QIAN Y., ITOH T., "Progress in active integrated antennas and their applications", *IEEE Transactions on MTT*, vol. 46, no.11, pp. 1891–1990, 1998.

[QUB 02] QUB, http://www.ee.qub.ac.uk/hfe/st173.htm.

[ROB 12] ROBERG M., FALKENSTEIN E., POPOVIC Z., "High-efficiency harmonically-terminated rectifier for wireless powering applications", *Proceedings of the International Microwave Symposium (IMS), IEEE,* TH1A-4, 2012.

[RUI 12] RUIZ M.N., MARANTE R., GARCÍA J.A., "A class E synchronous rectifier based on an E-pHEMT device for wireless powering applications", *Proceedings of the International Microwave Symposium (IMS)*, TH1A-3, 2012.

[SAE 11] SAEN T., ITOH K., BETSUDAN S., *et al.*, "Fundamentals of the bridge RF rectifier with an impedance transformer", *Proceedings of the 2011 IEEE MTT-S International Microwave Workshop Series on Innovative Wireless Power Transmission: Technologies, Systems, and Applications (IMWS-IWPT2011)*, pp. 255–258, 2011.

[SAK 97] SAKA T., FUJINO Y., FUJITA M., *et al.*, "An experiment of a C band rectenna", *Proceedings of SPS '97*, pp. 251–253, 1997.

[SAK 10] SAKAGUCHI K., WICAKSONO R.P., MIZUTANI K., *et al.*, Wireless grid to realize ubiquitous networks with wireless energy supply, IEICE Technical Report, vol. 109, no. 442, SR2009-113, pp. 149–154, 2010. [In Japanese]

[SAM 09] SAMPLE A.P., SMITH J.R., "Experimental results with two wireless power transfer systems", *Proceedings of the 2009 IEEE Radio & Wireless Symposium (RWS)*, pp. 16–18, 2009.

[SAN 00] SANG L.C.K., CELESTE A., LUK J.-D.L.S., "A point-to-point terrestrial wireless power transportation using an injection-locked magnetron array", *Proceedings of the Millennium Conference on Antennas & Propagation*, p. 387, 2000.

[SAS 04] SASAKI S., TANAKA K., KAWASAKI S., *et al.*, USEF SPS STUDY TEAM, "Conceptual study of SPS demonstration experiment", *The Radio Science Bulletin*, no. 310, pp. 9–14, 2004.

[SCH 88] SCHLESAK J.J., ALDEN A., OHNO T., "A microwave powered high altitude platform", *IEEE MTT-S International Symposium Digest*, pp. 283–286, 1988.

[SEB 01] SEBOLDT W., KLIMKE M., LEIPOLD M., *et al.*, "European sail tower SPS concept", *Acta Astronautica*, vol. 48, nos. 5ron, pp. 785.ona, 2001.

[SEB 04] SEBOLDT W., "Space- and earth-based solar power for the growing energy needs of future generations", *Acta Astronautica*, vol. 55, nos. 3 Astro 389 Astronaut.

[SHA 88] SHARP, http://friendsofcrc.ca/SHARP/sharp.html, 1988.

[SHI 97] SHIBATA T., AOKI Y., OTSUKA M., *et al.*, "Microwave energy transmission system for microrobot", *IEICE Transactions on Electronics*, vol. E80-C, no.2, pp. 303–308, 1997.

[SHI 98a] SHINOHARA N., MATSUMOTO H., "Dependence of DC output of a rectenna array on the method of interconnection of its array element", *Electrical Engineering in Japan*, vol. 125, no.1, pp. 9–17, 1998.

[SHI 98b] SHINOHARA N., MATSUMOTO H., "Experimental study of large rectenna array for microwave energy transmission", *IEEE Transactions on MTT*, vol. 46, no. 3, pp. 261–268, 1998.

[SHI 00] SHINOHARA N., FUJIWARA J., MATSUMOTO H., "Development of active phased array with phase-controlled magnetrons", *Proceedings of ISAP 2000*, vol. 2, pp. 713–716, 2000.

[SHI 01] SHINOHARA N., MITANI T., MATSUMOTO H., "Development of phase-controlled magnetron", *IEICE Transactions C*, vol. J84-C, no. 3, pp. 199–206, 2001. [In Japanese]

[SHI 03] SHINOHARA N., MATSUMOTO H., HASHIMOTO K., "Solar power station/satellite (SPS) with phase controlled magnetrons", *IEICE Transactions on Electronics*, vol. E86-C, no. 8, pp. 1550–1555, 2003.

[SHI 04a] SHINOHARA N., MATSUMOTO H., YAMAMOTO A., *et al.*, Development of high efficiency rectenna at mW input, IEICE Technical Report, SPS2004-08, pp. 15–20, 2004. [In Japanese]

[SHI 04b] SHINOHARA N., MATSUMOTO H., HASHIMOTO K., "Phase-controlled magnetron development for SPORTS: space power radio transmission system", *The Radio Science Bulletin*, no. 310, pp. 29–35, 2004.

[SHI 04c] SHINOHARA N., MATSUMOTO H., "Wireless charging system by microwave power transmission for electric motor vehicles", *IEICE Trans. C*, vol. J87-C, no. 5, pp. 433–443, 2004. [In Japanese]

[SHI 04d] SHINOHARA N., MATSUMOTO H., MITANI T., *et al.*, Experimental study on 'wireless power space', IEICE Technical Report, SPS2003-18, pp. 47–53, 2004. [In Japanese]

[SHI 05] SHINOHARA N., MITANI T., MATSUMOTO H., "Study on ubiquitous power source with microwave power transmission", *Proceedings of the International Union of Radio Science (URSI) General Assembly 2005*, CD-ROM C07.5(01145).pdf, 2005.

[SHI 07a] SHINODA H., MAKINO Y., YAMAHIRA N., *et al.*, "Surface sensor network using inductive signal transmission layer", *Proceedings of the International Conference on Networked Sensing Systems (INSS) 2007*, pp. 201–206, 2007.

[SHI 07b] SHINOHARA N., NAGANO K., ISHII T., *et al.*, "Experiment of microwave power transmission to the moving rover", *Proceeding of International Symposium on Antennas and Propagation (ISAP2007)*, *IEICE*, 3B1-1, 2007.

[SHI 08] SHINOHARA N., MIYATA Y., MITANI T., *et al.*, "New application of microwave power transmission for wireless power distribution system in buildings", *Proceedings of Asia-Pacific Microwave Conference 2008*, *IEICE*, CD-ROM H2-08.pdf, 2008.

[SHI 11a] SHINOHARA N., "Beam efficiency of wireless power transmission via radio waves from short range to long range", *Journal of the Korean Institute of Electromagnetic Engineering and Science*, vol. 10, no. 4, pp. 224–230, 2011.

[SHI 11b] SHINOHARA N., "Wireless charging system of electric vehicle with GaN Schottky diodes", *Proceedings of the International Microwave Symposium (IMS) Workshops, WFA: Wireless Power Transmission*, Baltimore, 2011.

[SHI 12a] SHINOHARA N., "Recent wireless power transmission via microwave and millimeter-wave in Japan", *42nd European Microwave Conference*, pp. 1347–1350, 2012.

[SHI 12b] SHINOHARA N. (ed.), *Solar Power Satellite/Station*, Ohmsha Publishing, 2012. [In Japanese]

[SHI 13a] SHINOHARA N., "Wireless power transmission progress for electric vehicle in Japan", *Proceedings of the 2013 IEEE Radio & Wireless Symposium (RWS)*, pp. 109–111, 2013.

[SHI 13b] SHINOHARA N., "Wireless power transmission in URSI", *Proceedings of the 2nd International Conference on Telecommunications and Remote Sensing (ICTRS2013)*, pp. 49–53, 2013.

[SHI 13c] SHINOHARA N., "Beam control technologies with a high-efficiency phased array for microwave power transmission in Japan", *Proceedings of IEEE*, vol. 101, no. 6, pp. 1448–1463, 2013.

[SHI 13d] SHINOHARA N., YUTA KUBO, "Suppression of unexpected radiation from microwave power transmission system toward electric vehicle", *Proceedings of the 2013 Asia-Pacific Radio Science Conference (AP-RASC)*, E3-4 (No. 290450), 2013.

[SHI 13e] SHINOHARA N., KUBO Y., TONOMURA H., "Mid-distance wireless power transmission for electric truck via microwaves", *Proceedings of the 2013 International Symposium on Electromagnetic Theory (EMT-S2013), URSI*, pp. 841–843, 2013.

[SHO 13] SHOWA AIRCRAFT, http://www.showa-aircraft.co.jp/products/EV/ kyuuden.html, 2013. [In Japanese]

[SIV 94] SIVAN L., *Microwave Tube Transmitters – Microwave Technology Series 9*, Chapman & Hall, 1994.

[SKO 90] SKOLNIK M.I., *Radar Handbook*, 2nd ed., McGraw-Hill, 1990.

[SMI 10] SMITH J.R., "Mapping the space of wirelessly powered systems", *Proceedings of the International Microwave Symposium (IMS) Workshops, IEEE*, WFB-3, 2010.

[SPS 93] SPS 2000 TASK TEAM, "SPS 2000 Project Concept – a strawman SPS system", S2-T1-X, Preliminary, July 1993. [In Japanese, English summary is available]

[STA 74] STARK L., "Microwave theory of phased array antenna – a review", *Proceedings of IEEE*, vol. 62, pp. 1661–1701, 1974.

[STE 87] STEPHAN K.D., MORGAN W.A., "Analysis of interinjection-locked oscillators for integrated phased arrays", *IEEE Transactions on AP*, vol. AP-35, pp. 771–781, 1987.

[SUH 02] SUH Y.H., CHANG K., "A high-efficiency dual-frequency rectenna for 2.45- and 5.8-GHz wireless power transmission", *IEEE Transactions on MTT*, vol. 50, no.7, pp. 1784–1789, 2002.

[SUM 03] SUMMERER L., ONGARO F., VASILE M., *et al.*, "Prospects for space solar power work in Europe", *Acta Astronautica*, vol. 53, nos. 4–10, pp. 571–575, 2003.

[SUN 03] SUNG S.S., ROQUE J.D., MURAKAMI B.T., *et al.*, "Retrodirective antenna technology for CubeSat networks", *Proceedings of the IEEE Topical Conference on Wireless Communication Technology 2003*, pp. 220–221, 2003.

[SYS 94] SYSTEMS CONTROL TECHNOLOGY, INC., Roadway powered electric vehicle project rack construction and testing program phase 3D, California PATH Research Paper UCB-ITS-PRR-94-07, ISSN 10551425, March 1994.

[TAH 05] TAHIR I., DEXTER A., CARTER R., "Phase locked magnetrons by use of their pushing characteristics", *Proceedings of the 6th International Vacuum Electronics Conference (IVEC2005)*, pp. 65–68, 2005.

[TAH 06] TAHIR I, DEXTER A., CARTER R., "Frequency and phase modulation performance of an injection-locked CW magnetron", *IEEE Transactions on ED*, vol. 53, no.7, pp. 1721–1729, 2006.

[TAK 05] TAKADA Y., *et al.*, "C-band AlGaN/GaN HEMTs with 170W output power", *Proceedings of the 2005 International Conference on Solid State Devices and Materials*, Extended Abstracts, pp. 486–487, September 2005.

[TAK 06] TAKAGI K., *et al.*, "X-band AlGaN/GaN HEMT with over 80W output power", *Proceedings of the 28th IEEE Compound Semiconductor IC Symposium*, N.5, November 2006.

[TAK 09] TAKAHASHI K., AO J.-P., IKAWA Y., *et al.*, "GaN Schottky diodes for microwave power rectification", *Japanese Journal of Applied Physics*, vol. 48, no. 4, pp. 04C095-1–04C095-4, 2009.

[TAK 13] TAKACS A., AUBERT H., DESPOISSE L., FREDON S., "K-band energy harvesting for satellite application", *Proceedings of the International Microwave Symposium (IMS)*, WE3G-1, 2013.

[TES 04a] TESLA N., "The transmission of electric energy without wires", *The 13th Anniversary Number of the Electrical World and Engineer*, 5 March 1904.

[TES 04b] TESLA N., *Experiments with Alternate Current of High Potential and High Frequency*, McGraw Publishing Company, New York, 1904.

[TRE 02] TREW R.J., "SiC and GaN transistors – is there one winner for microwave power applications?", *Proceedings of IEEE*, vol. 90, no.6, pp. 1032–1047, 2002.

[TYL 58] TYLER V.J., "A new high-efficiency high power amplifier", *Marconi Review*, vol. 21, no. 130, pp. 96–109, Fall 1958.

[UED 04] UEDA Y., FUJIMORI K., NOGI S., *et al.*, "Efficient conversions of rectification circuit for low power rectennas", *Proceedings of IEICE*, SBC-1-9, 2004. [In Japanese]

[URS 07] URSI INTER-COMMISSION WORKING GROUP ON SPS, URSI white paper on solar power satellite (SPS) systems and report of the URSI inter-commission working group on SPS, June 2007.

[VAN 59] VAN ATTA, L.G., "Electromagnetic reflector", U.S. Patent No. 2,908,002, 6 October 1959.

[VAN 82] VANKE V.A., LOPUKHIN V.M., ROSNOVSKY V.K., *et al*,, "Ground-based receiving/converting system for space solar power systems", *Radiotechnique & Electronics*, vol. 27, no. 5, p. 1014, 1982.

[VAN 91] VANKE V.A., SAVVIN V.L., "Cyclotron wave converter for SPS energy transmission system", *Proceedings of SPS '91*, pp. 515–520, 1991.

[VAN 95] VANKE V.A., Savvin V.L., BOUDZINSKI I.A., *et al.*, "Development of cyclotron wave converter", *Proceedings of WPT '95*, p. 3, 1995.

[VAN 98] VANKE V.A., MATSUMOTO H., SHINOHARA N., KITA A., "Cyclotron wave converter of microwave into DC", *IEICE Transactions on Electronics*, vol. E81-C, no. 7, pp. 1136–1142, 1998.

[VAN 03] VANKE V.A., MATSUMOTO H., SHINOHARA N., "On a possibility to decrease magnetic intensity in microwave/DC cyclotron wave converter", *IEICE Transactions on Electronics*, vol. E86-C, no.7, pp. 1390–1392, 2003.

[VYA 12] VYAS R.J., NISHIMOTO H., TENTZERIS M., *et al.*, "A battery-less, energy harvesting device for long range scavenging of wireless power from terrestrial TV broadcasts", *Proceedings of the International Microwave Symposium (IMS)*, IEEE, TU4F-2, 2012.

[WAK 00] WAKEJIMA A., *et al.*, "370W output power GaN-FET amplifier for W-CDMA cellar base stations", *Electronics Letters*, vol. 41, no. 25 pp. 1371–1372, December 2000.

[WAK 05] WAKEJIMA A., *et al.*, "280W output power single-ended amplifier using single-die GaN-FET for WCDMA cellar base stations", *Electronics Letters*, vol. 41, no. 18, pp. 1004–1005, September 2005.

[WAN 13] WANG X., MORTAZAWI A., "A self-sensing AM frequency electromagnetic energy scavenger", *Proceedings of the International Microwave Symposium (IMS)*, WE4G-4, 2013.

[WAT 66] WATSON D.C., TABBOT K.T., JONSON C.C., "A cyclotron wave microwave power converter", *Proceedings of IEEE*, no. 11, p. 1797, 1966.

[WAT 68] WATSON D.C., GROW R.W., JONSON C.C., "A rectifier with transverse interaction", in OKRESS E.C. (ed.), *Microwave Power Engineering*, vol. 1, Academic Press, New York/London, pp. 408–419, 1968.

[WAT 70] WATSON D.C., GROW R.W., JONSON C.C., "A cyclotron wave rectifier for S-band and X-band", *Journal of Microwave Power*, vol. 5, no. 2, p. 72, 1970.

[WIP 13] WIPOT, http://www.wipot.jp/english/index.html, 2013.

[WPC 13] WPC, http://www.wirelesspowerconsortium.com/, 2013.

[WPM 13] WPMC, http://wpm-c.com/, 2013. [In Japanese]

[YAM 92] YAMAGIWA Y., NAGATOMO M., "An evaluation model of solar power satellite using world dynamics simulation", *Space Power*, vol. 11, no. 2, pp. 121–131, 1992.

[YAM 05] YAMANAKA K., *et al.*, "S and C band over 100W GaN HEMT 1chip high power amplifiers with cell division configuration," *Proceedings of the 2005 European Gallium Arsenide and Other Semiconductor Application Symposium*, pp. 241–244, October 2005.

[YAM 07a] YAMAGATA S. TANAKA, SHOGEN K., "Broadcasting satellite system using onboard phased array antenna in 21-GHz band", *Proceedings of the 8th International Vacuum Electronics Conference (IVEC2007)*, pp. 209–210, 2007.

[YAM 07b] YAMANAKA K., *et al.*, "C-band GaN HEMT power amplifier with 220W output power", *Proceedings of the International Microwave Symposium*, 3TH1A-02, June 2007.

[YAM 10] YAMANAKA K., TUYAMA Y., OHTSUKA H., *et al.*, "Internally-matched GaN HEMT high efficiency power amplifier for space solar power stations", *Proceedings of the Asia-Pacific Microwave Conference 2010*, pp. 119–122, 2010.

[YAM 13] YAMASHITA T., HONDA K., OGAWA K., "High efficiency MW-Band rectenna using a coaxial dielectric resonator and distributed capacitors", *Proceedings of the 2013 International Symposium on Electromagnetic Theory (EMTS2013)*, pp. 823–826, 2013.

[YAN 13] YANG C.L., YANG Y.-L., YANG C.-W., "Adaptive pulse waveform modulation to enhance wireless power efficiency for biomedical applications", *Proceedings of the International Microwave Symposium (IMS)*, WE3G-3, 2013.

[YOO 92] YOO T.W., CHANG K., "Theoretical and experimental development of 10 and 35 GHz rectennas", *IEEE Transactions on MTT*, vol. 40, no. 6, pp. 1259–1266, 1992.

[YOR 98] YORK R.A., ITOH T., "Injection- and phase-locking techniques for beam control", *IEEE Transactions on MTT*, vol. 46, no. 11, pp. 1920–1929, 1998.

[YOS 09] YOSHIOKA K., MATSUOKA H., HAYAMI H., *Essays on the Solar Power Satellite: An Interdisciplinary Study of Space Industry and the Future of Society,* Keio University Publishing, 2009. [In Japanese]

[ZBI 06] ZBITOU J., LATRACH M., TOUTAIN S.,"Rectenna and monolithic integrated zero-bias microwave rectifier", *IEEE Transactions on MTT*, vol. 54, no. 1, pp. 147–152, 2006.

[ZEL 82] ZELL C.E., BOLGER J.G., "Development of an engineering prototype of a roadway powered electric transit vehicle system," *Proceedings of the 32nd IEEE Vehicular Technology Conference*, vol. 32, pp. 435–438, May 1982.

[ZHE 07] ZHENG C., YOSHIDA T., ISHIKAWA R., *et al.*, "GaN HEMT class-F amplifier operating at 1.9GHz", *Proceedings of IEICE*, C-2-27, March 2007. [In Japanese]

[ZHU 11] ZHU N., ZIOLKOWSKI R.W., XIN H., "Electrically small GPS L1 rectennas", *IEEE Antennas Wireless Propagation Letters*, vol. 10, pp. 935–938, 2011.

Index

O, P, Q

R, S, T

U, V, W

Printed and bound by CPI Group (UK) Ltd, Croydon, CR0 4YY
13/09/2026
14961078-0003